AF700353

Food Defense Fundamentals

Food Defense Fundamentals

Using the SHARE Principle to Protect The Global Food Supply

Michael Dixon
Ph.D, LDN, CNS

Tara Paster
CP-FS, *FMP*

PEARSON

Boston Columbus Indianapolis New York San Francisco Upper Saddle River
Amsterdam Cape Town Dubai London Madrid Milan Munich Paris Montreal Toronto
Delhi Mexico City Sao Paulo Sydney Hong Kong Seoul Singapore Taipei Tokyo

Editorial Director: Vernon Anthony
Senior Acquisitions Editor: William Lawrensen
Editorial Project Manager: Tracey Dills
Editorial Assistant: Lara Dimmick
Director of Marketing: David Gesell
Senior Marketing Manager: Thomas Hayward
Marketing Assistant: Les Roberts
Associate Managing Editor: Alexandrina Benedicto Wolf
Production Editor: Kris Roach
Production Manager: Susan Hannahs
Senior Art Director: Jayne Conte
Cover Designer: Bruce Kenselaar
Cover Art: Shutterstock
Full-Service Project Management: George Jacob/Integra Software Services
Composition: Integra Software Services Pvt. Ltd.
Text Printer/Bindery: Bind-Rite Graphics, Inc.
Cover Printer: Lehigh-Phoenix Color
Text Font: Minion 11/13

Credits and acknowledgments borrowed from other sources and reproduced, with permission, in this textbook appear on the appropriate page within the text.

Photo Credits: Shutterstock, pp. 3,4,36,46,49,57,63,65,70,74,75,79; Thinkstock, pp. 6,98; Dreamstime LLC, pp. 7,11,21,26,34; Fotolia LLC, pp. 20,27,28,48; Alamy –UK & India pp. 25,29,67; iStockphoto.com, pp. 28,30,71; AP Images, pp.99; NewsCom, pp. 25,26,30; National Pasteurized Eggs, pp. 35

Copyright © 2012 by Pearson Education, Inc. All rights reserved. Printed in the United States of America. This publication is protected by Copyright and permission should be obtained from the publisher prior to any prohibited reproduction, storage in a retrieval system, or transmission in any form or by any means, electronic, mechanical, photocopying, recording, or likewise. To obtain permission(s) to use material from this work, please submit a written request to Pearson Education, Inc., Permissions Department, One Lake Street, Upper Saddle River, New Jersey 07458 or you may fax your request to 201-236-3290.

Many of the designations by manufacturers and seller to distinguish their products are claimed as trademarks. Where those designations appear in this book, and the publisher was aware of a trademark claim, the designations have been printed in initial caps or all caps.

Library of Congress Cataloging-in-Publication Data
Dixon, Michael.
Food defense fundamentals : using the S.H.A.R.E. principle to protect the global food supply / Michael Dixon, Tara Paster.
p. cm.
Includes index.
ISBN-13: 978-0-13-139181-9
ISBN-10: 0-13-139181-X
1. Food—Safety measures. 2. Food contamination—Prevention. 3. Bioterrorism—Prevention. 4. Food industry and trade—Safety measures. I. Paster, Tara, 1968- II. Title.
RA601.D63 2012
363.19'26—dc23

2011025200

10 9 8 7 6 5 4 3 2 1

PEARSON

ISBN 10: 0-13-139181-X
ISBN 13: 978-0-13-139181-9

CONTENTS

Preface ix

Chapter 1 **Strategy 1**

Overview 2

Learning Objectives 2

Exercise: What Is Your Strategy? 2

Reality Check 4

Key Terms Match Game 4

Leadership, Safety, and Security Protocols 5

What's Next? Where to get started? 5

How Detailed is your Strategic Plan 6

How to Evaluate Your Plans 6

What should be in a Food Defense SHARE plan? 7

Human Resource Systems 8

■ Employee Awareness SOP 9

How do you create a Food Defense Culture That is Conscientious, But Not Alarming? 9

Human Resource Food Defense Audit Protocol 10

Manager's Food Defense Audit Protocol 10

Employee's Food Defense Audit Protocol 11

Physical Security Food Defense Audit Protocol 11

Recognize the Importance of Food Defense Strategy 12

Strategy Study Questions 12

■ Apply Your Knowledge (Close Hold Assignment) 15

Chapter 2 **Hazards 17**

Overview 18

Learning Objectives 18

Exercise: What Are Your Vulnerabilities? 19

Understanding Vulnerability 20

Reality Check 21

Key Terms Match Game 22

Leadership, Safety, and Security Protocols 23

What's Next? Where to Get Started with Hazards? 23

Food Defense and the Global Food Supply 23

Tourist Population: Airports, Cruise Ships, Hotels, and Casinos 24

AIRPORTS 24

CRUISE SHIPS 25

HOTELS 26

CASINOS 26

Colleges/Universities 26

COLLEGES 26

At-Risk Populations: Hospitals, Schools, and Senior Care Facilities 27
Hospitals 27
Schools 27
Senior Care Facilities 28
Quick Service, Casual Dining, Fine Dining, Food Service Operations: Restaurants 28
Restaurants 28
Retail Food Outlets: Convenience Stores, Grocery Stores, and Supermarkets 29
Convenience Stores 29
Grocery Stores 29
Government: Commissaries and Military Restaurants 29
CARVER+Shock Methodology 30
CARVER+Shock Weaknesses 32
What Is the HACCP Star? 32
Food Defense and HACCP Principle 1: Conduct a Hazard Analysis 33
Food Defense and HACCP Principle 2: Determine Critical Control Point 34
Food Defense and HACCP Principle 3: Establish Critical Limits 34
Food Defense and HACCP Principle 4: Monitoring 34
Food Defense and HACCP Principle 5: Taking Corrective Actions 36
Food Defense and HACCP Principle 6: Verification 37
Food Defense and HACCP Principle 7: RecordKeeping and Documentation 38
National Center for Food Protection and Defense 38
Hazards Study Questions 39
■ Apply Your Knowledge (Close Hold Assignment) 42

Chapter 3 Accessibility 44
Overview 45
Learning Objectives 45
What Do You control? What Do You Influence? 45
Reality Check 48
Key Terms Match Game 49
Leadership, Safety, and Security Protocols 50
What's Next? Where to Get Started with Accessibility? 50
Food Defense Share Accessibility Levels 51
Open—Easily Accessible 51
High—Accessible 52
Medium—Partially Accessible 52
Low—Hardly Accessible 53

Closed—Not Accessible 53
Apply Motive and Accessibility to the Reality Check 54
Accessibility: A Shared Responsibility 56
Accessibility Study Questions 58
■ Apply Your Knowledge (Close Hold Assignment) 61

Chapter 4 Restriction 63
Overview 64
Learning Objectives 64
Exercise: What Is Restriction? 64
Reality Check 66
Key Term Match Game 68
Leadership, Safety, and Security Protocols 68
What's Next? Where to Get Started? 68
A Functional Food Defense Plan 84
What Should Be in a Food Defense SHARE Plan? 84
Now, It Is Time to Create a Food Defense SHARE Plan 86
Restriction Study Questions 91
■ Apply Your Knowledge (Close Hold Assignment) 94

Chapter 5 Enforcement 96
Overview 97
Learning Objectives 97
Exercise: What Is Enforcement? 97
What Are Your Goals? 97
Exercise Summary 98
Reality Check 99
Key Term Match Game 101
Leadership, Safety, and Security Protocols 102
What's Next? Where to Get Started? 102
Credibility of Evidence 103
Criminal Investigation Process 103
Critical Information 103
Threat Analysis 104
Law Enforcement Investigation Goals 104
Protect the Public Safety 104
Prevent a Criminal Act 104
Identify, Apprehend, and Prosecute the Perpetrators 104
Protect Law Enforcement Personnel 104
Data Collection and Law Enforcement 105
FDA/USDA Decontamination and Disposal Guidance to Industry: Characteristics of an Intentional Adulteration Incident 105
Procedures for Incidents Involving Intentional Adulteration 106

Voluntary National Retail Food Regulatory Program Standards—April 2009—Standard 5 106

Final Thought Questions? 110

Enforcement Study Questions 112

■ Apply Your Knowledge (Close Hold Assignment) 114

Appendix 1: *Food Defense Plan: Security Measures for Food Defense 116*

Appendix 2: *Business Emergency Plan 127*

Glossary 134

Index 137

PREFACE

Our mission is to provide a comprehensive global food defense solution that protects our food supply from acts of deliberate contamination or intentional adulteration, ultimately protecting public health and the economy. Our goal is to increase food defense knowledge, skills, and ability by providing the means to accomplish this objective. Rapid worldwide communication and expanding global food markets provide the driving force behind every segment of our food supply and result in a food defense culture change. This food defense training program is designed to help achieve this goal. Food defense enables every food industry sector and every person involved to protect public health by ensuring the safety of the food supply.

- This manual is the highest-level training course designed for managers, employees, and leaders on how to create a food defense culture, write a food defense plan, and build the confidence of managers and employees as a result of this course.
- This manual includes knowing the differences between food safety and food defense. We shall discuss the concepts of retail foodservice in terms of the flow of food and how food defense is essential to your daily activities.

Food defense is critical to the safety of our world. All parties from farm to fork must contribute to the protection of the global food supply. Global challenges are many, such as tough economic times, environmental disasters, fuel supply issues, terrorist activity, and even a worldwide pandemic. Remember, a food defense oversight could lead to disaster. This is why industry and regulators must be collaborative partners in establishing and implementing food defense practices. It is essential to acknowledge the fact that food defense is forever changing. That is why the culture of your organization must be vigilant in maintaining food defense standard operating procedures.

Following is a list of challenges typically faced when adding food defense standard operating procedures into a business and/or regulatory culture:

- Limited comprehensive training programs for managers and employees
- Incorporating food defense lessons learned (also called best practices) into human resource policies and procedures
- Deficiency in overall awareness of food defense and food safety
- Financial impact, such as the return on investment, is sometimes viewed as just another line item on the profit and loss statement
- Regulatory authorities require controls to be implemented to address the risk to public safety
- Workforce issues like overcoming high employee turnover, language barriers, and varied educational levels
- Complex systems that include a variety of menu items with frequently changing ingredients and procedures
- Retail foodservice operator's inability to train situation awareness and maintain vigilant behaviors

The SHARE (*S*trategy, *H*azards, *A*ccessibility, *R*estrictions, and *E*nforcement) principles provide a solid foundation for you to make the plan to protect against the intentional adulteration of food. SHARE principles include the incorporation of science-based strategies to prepare and protect the food supply chain at specific vulnerable points in your operation. SHARE principles focus on maintaining food safety and deterring food defense problems. In addition, rather than relying primarily on reacting to problems after they occur Food Defense SHARE will guide you in the right direction for success.

This certification program was designed to be easily customized for convenience stores, supermarkets, restaurants, casinos, military, schools, senior care facilities, airports, institutional foodservice, cruise ships, contract foodservice, hotels, commissaries, hospitals, distribution centers, colleges, and universities. The focus is on the need for these industry sectors to realize the threats to the food chain.

In the following chapters you'll be exposed to concepts that are designed to help you in your day-to-day operations, such as monitoring key positions within the flow of food, relating to key concepts in Hazard Analysis and Critical Control Point (HACCP) and how they provide another layer of protection for the food that you serve. Finally, we shall look at bad practices, or habits, in the foodservice industry and how SHARE principles will enable and allow you to adopt the best principles of food defense for your immediate span of control.

To provide thorough food defense is to SHARE the responsibility throughout your organization. SHARE is our acronym for *S*trategy, *H*azards, *A*ccessibility, *R*estriction, and *E*nforcement [EI5]to respond effectively when problems emerge despite preventive controls. Using SHARE will direct reliance on vulnerability assessments, self-inspections, third-party audits, training, and human resource systems to meet increased food defense mandates for both domestic and foreign facilities. Following is a brief description of each chapter in *SHARE*:

- In Chapter 1, we will start with *S*trategy. Strategy requires creating a *culture of commitment* throughout the organization where each individual does his or her best to ensure safe food. This allows the teams to be at their best, which allows the organization to be at its best and allows consumers to have confidence in your brand or agency.
- In Chapter 2, *H*azards, you will understand the potential negative impact caused by the introduction of hazards in the flow of food. A food defense SHARE plan must address vulnerabilities and hazards. If vulnerabilities and hazards are not appropriately identified, you will not be able to prevent, eliminate, and reduce risks associated with an intentional adulteration event, which is an act of terrorism.
- Chapter 3, *A*ccessibility, discusses your opportunity to limit an insider attacker's ability to wreak havoc because you are directing responsibilities, assignments, and timelines. Accessibility rights need to be determined for each aspect of the flow of food: purchase, receive, store, prepare, cook, hold, cool, reheat, and serve. Any of these steps in the flow of food are potential entry points for hazards to intentionally be introduced into the product or processing stream.
- Chapter 4, *R*estriction, focuses on having a well-defined physical security protocol and incorporating clear job descriptions with vigilant oversight within the boundaries of your operation. Restriction must be used as a proactive and preventive measure to defend against your vulnerabilities and guard your *brand's shock value (BSV)*.
- Finally, in Chapter 5, we'll consider *E*nforcement and how you can appropriately *respond to an alleged attack* at your facility. Enforcement is traditionally defined as legal actions taken to gain compliance with laws, rules, regulations, and agreements, resulting in penalties or criminal sanctions for violations.

Your company is safer because you are here engaging in the fight to protect the clients and customers that you serve directly and we serve collectively. Thanks for joining us on the journey to a more empowered society, a society that is prepared and not scared.

We would like to express our thanks to the following reviewers for their insightful feedback, support, and help throughout developing this book:
List of Reviewers:

- Bharat L. Patel, DVM (Public Health Veterinarian - Retired USDA)
- David W. K. Acheson, MD (Managing Director, Food and Import Safety Practice - Leavitt Partners)
- Theresa Stretch, MS, RD (Registered Dietitian - University of Mississippi)
- David Oryang, MS (Mathematical Statistician/Risk Analyst - FDA)
- Fay Algeo (Senior Training Specialist, Paster Training, Inc.)
- George Zameska, MS (Vice President of Regulatory Affairs, Paster Training, Inc.)
- Susan Algeo, MPH (Vice President of Training, Paster Training, Inc.)

List of Class Reviewers from Rutter's Farm Stores:

- Jere Matthews - Vice President, Operations
- Deb Young - Training Supervisor
- Chuck Moyer - Training Supervisor
- Jo Smith - Food Service Quality Assurance Supervisor
- Mindy Torney - Food Service Quality Assurance Supervisor

Add Disclaimer:

- Dr. Michael O. Dixon is currently serving as a Policy Analyst with the U.S. Food and Drug Administration; and, the views expressed in the book are those of the author and do not necessarily reflect the views of the FDA or of the Federal Government.

Food Defense Fundamentals

Chapter 1
Strategy

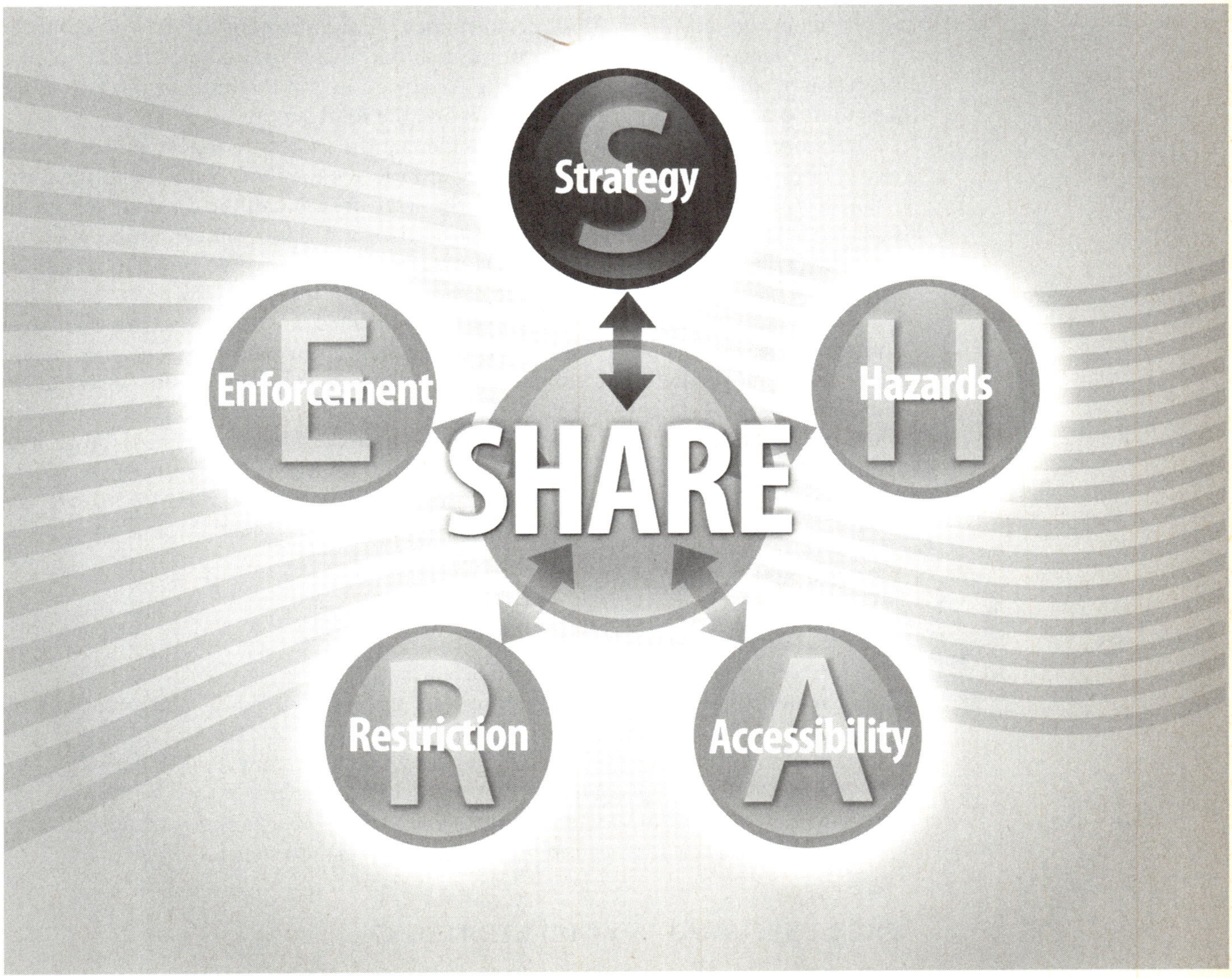

OVERVIEW

SHARE is an acronym of the key food defense components—strategy, hazards, accessibility, restriction, and enforcement.

Strategy requires creating a culture of **commitment** through all levels of an organization to prevent any unintentional (food safety) or intentional (food defense) contamination event.

SHARE is an acronym of the key food defense components—strategy, hazards, accessibility, restriction, and enforcement. Strategy represents the fundamental core of protecting our food supply, and that is why it is a nonnegotiable component of any plan of action. Strategy requires creating a **culture of commitment** through all levels of an organization to prevent any unintentional (food safety) or intentional (food defense) contamination event.

For this goal to be met, the strategy must engage all people and tasks involved in the flow of food. The strength of any strategy will depend on a committed work ethic that builds a **supportive culture** needed to remove **barriers** and improve food safety and food defense. Your organization's operation should be similar to the operation of a precision machine. If you examine a precision machine, there are key factors such as efficiency, design, innovation, performance, and craftsmanship. As you consider your food defense culture, keep in mind it should incorporate the components of a precision machine. Strategy involves the construction and assembly of infrastructure intended to accomplish the goal of protecting the food supply.

An organization's commitment to achieving the successful outcome of safe food works best when each member of the organization is totally committed to the same goals. If you do not do your job right, the next person cannot do their job right. Likewise, if the person before you did not do their job right, you cannot do your job right. This dependency builds a **culture** of **interdependency**. No one person can be isolated; management must SHARE the responsibility from farm to fork.

In addition, if your food defense plan is mishandled, it is a recipe for disaster. Most companies treat their food defense plan as **proprietary information**, that is, information that needs to be secured. Therefore, careful guidance in the **vulnerability assessments** and **close hold** assignments within the Food Defense SHARE program is recommended ultimately to protect your employees, customers, brand, and industry. Here is the padlock icon that represents confidential information that should not be disclosed. Now, let's begin with the learning objectives of the first step in Food Defense SHARE for strategy.

LEARNING OBJECTIVES

In this chapter, you will learn to:

- Understand strategy as it relates to SHARE.
- Create a food defense culture for your operation using a systematic human resource approach.
- Strengthen your food defense culture and set the right tone.
- Explain how concepts of leadership extend to food defense.
- Identify your role in food defense as a conscientious leader and/or manager.
- Determine a communication protocol for information sharing.

EXERCISE: WHAT IS YOUR STRATEGY?

This includes having a food defense culture, dedicated leadership, established human resource protocols, and effective training for managers and employees.

Strategy involves setting the right tone to achieve food defense awareness throughout your operation. This includes having a food defense culture, dedicated leadership, established human resource protocols, and effective training for managers and employees. The exercise Strategy Vulnerability Assessment, given in the following box, will help you to determine if your organization's strategy is up-to-date as it relates to protecting the food supply. Answer *yes* or *no* to the following questions about your operation. Place a check in the appropriate box. You will use this assessment later when you complete your Food Defense SHARE plan.

Figure 1.1 Culture of Interdependency.

Strategy Vulnerability Assessment	**Yes**	**No**
1. Does management recognize the importance of food defense?	☐	☐
2. Is food defense a priority to management?	☐	☐
3. Do all employees know the food defense mission?	☐	☐
4. Do all employees support the mission of food defense?	☐	☐
5. Are procedures established with key input from employees?	☐	☐
6. Can employees get help from managers if needed?	☐	☐
7. Can managers get help from employees if needed?	☐	☐
8. Are employees disciplined when they are not following procedures?	☐	☐
9. Are employees rewarded by management?	☐	☐
10. Do you challenge your managers to do what is required?	☐	☐
11. Are your leaders, managers, and employees trained in food safety and food defense?	☐	☐
12. Does your organization set the right tone by having food defense in your orientation process?	☐	☐
13. Does your organization set the right tone by having food defense in all job descriptions?	☐	☐
14. Does management offer a variety of learning opportunities in food safety and food defense?	☐	☐
15. Is your organization proactive in promoting food safety and food defense?	☐	☐

- If you answered *yes* to *all* of the questions, congratulations—this is proof that your organization is focused on achieving food defense strategies.
- However, if you answered *no* to *any* of the questions, it is recommended that your organization develop or update their food defense strategies and plan.

Reality Check

School Students Threatened by Contaminated Ground Beef

In a 2009 incident, the U.S. Department of Agriculture (USDA) recalled over 800,000 pounds of ground beef. Across eleven states, thirty-nine people reported getting sick from a dangerous strain of ***Salmonella*** that was discovered in the recalled ground beef. This type of *Salmonella*, which can cause diarrhea, abdominal cramps, fever, vomiting, and even death, was particularly resistant to antibiotics.

Surprisingly, the company that recalled the product was a major supplier of ground beef for the National School Lunch Program (NSLP); but the recall did not include schools. Though safety experts later reported that the beef should have been included in the recall, they also indicated that the tests the government used, at the time, were both inconsistent and often wrong. In fact, the recall was issued after the beef had already shipped to schools. The U.S. government has a "zero-tolerance" policy regarding *Salmonella* and ***E. coli*** O157:H7 in ground beef that is sold to the school lunch program. The zero-tolerance policy was put in place for the protection of school children.

If this were an intentional act of food terrorism, it could have resulted in deaths across multiple jurisdictions. An intentional act of food terrorism may involve the use of a more toxic biological or chemical agent that would devastate your brand, potentially your entire industry, and as importantly devastate the **psyche** of employees and consumers. Therefore, becoming more vigilant every day and incorporating SHARE principles is essential for all organizations from farm to fork. Let's look at some key terms to help understand strategy and your role.

Figure 1.2 School children eating in cafeteria.

Key Terms Match Game

Answer the questions below by filling in the blank with the appropriate word that best fits the sentence as it relates to the strategies of your organization.

Human Resources	**Protection**
Training	**Concept**
Leadership	**Management**
Job Description	**Nonnegotiable**
Barriers	**Information Sharing**

1. ______________________________
is the name given to the document that describes the responsibilities and duties of a particular position.

2. ______________________________
is reducing or guarding against risk or guarding against.

3. ______________________________
is the overall management of individuals.

4. ______________________________ is organizational activities that involve specific tasks such as planning, organizing, and controlling to achieve goals.

5. ______________________________ is indisputable and anything other than that is not acceptable.

6. ______________________________ is the distribution of knowledge.

7. ______________________________ is acquiring knowledge, skills, and ability.

8. ______________________________ is simply a mental explanation or abstract idea.

9. ______________________________ are obstacles that must be managed.

10. ______________________________ is the ability to inspire others to achieve great things.

The concept of information sharing gained popularity as a result of the terrorist events of September 11, 2001. The need for information sharing and continued **resiliency** within your organization begins with management taking responsibility and accountability for creating a food defense culture. This is why strategy requires the recognition of barriers and nonnegotiable procedures to ensure that your organization's actions and activities provide for protecting your livelihood and brand. The management and training of the SHARE principles must be utilized throughout the entire organization to prevent, eliminate, and reduce risks.

The need for information sharing and continued resiliency within your organization begins with management taking responsibility and accountability for creating a food defense culture.

LEADERSHIP, SAFETY, AND SECURITY PROTOCOLS

What's Next? Where to Get Started?

The essential core of food defense strategy involves efforts and actions needed to prevent acts of food terrorism. This needs to be remembered, adopted, and practiced by supervisors, managers, directors, and team leaders. It is your responsibility to take proper strategic action in providing leadership in this important area. To do this, you need to identify your role in food defense as a conscientious leader and/or manager. Food defense expectations must be established by the management of your organization and should be defined in all job descriptions at every level. Food defense, food safety, and Hazard Analysis and Critical Control Point (HACCP) are not the same, but they have significant overlap.

- Food defense is related to the *intentional* contamination of the food supply.
- Food safety is related to the *unintentional* contamination of the food supply.
- HACCP is a *proactive management system* comprising of seven principles to prevent, eliminate, and reduce hazards to a safe level.

Although these terms are different, the overarching principle of protecting the safety and security of your product is the same. Your facility may want to share resources and assemble one team for HACCP and food defense, while other organizations may decide to assemble an HACCP team and a separate food defense team to distribute the

Figure 1.3 Leadership, and a combination of back-of-house, and front-of-house team members.

workload responsibilities. Your designated team, whether together or separate, should not be larger than six members (depending on the size of your establishment) and should always include a team leader.

HOW DETAILED IS YOUR STRATEGIC PLAN

For food defense to work, everyone must work as a team and each individual must do his or her part to stop any potential food contamination. All managers, team leaders, and employees must ensure the protection of human life and health, as well as the brand, by being sensible, while maintaining the practicality to conduct business as usual. The team leader needs to assign responsibilities, train employees, hire reputable **contractors** that meet specific qualifications, and ensure that the food defense initiatives are accomplished.

How to Evaluate Your Plans

Even though you may use the same team leader and/or members to properly execute your HACCP and food defense initiatives, you will still need two different plans to capture the variance of details.

When out-of-control procedures are identified, the necessary corrective actions, also known as countermeasures, can be taken to ensure the product is safe.

The leader needs to represent the team, manage objectively, and request financial and other resources needed for a successful HACCP plan and a food defense plan. Even though you may use the same team leader and/or members to properly execute your HACCP and food defense initiatives, you will still need two different plans to capture the variance of details. For example, your HACCP plan may include contacting the local health department in the event of a potential foodborne outbreak and your food defense plan may include contacting law enforcement in the suspicion of intentional product contamination.

Whether it is HACCP or Food Defense SHARE, continual self-inspection is needed to successfully evaluate both plans. The process of continual self-inspection develops employee and management behaviors instead of relying on a specific moment in time. On-site self-inspection should be used to alert management of out-of-control procedures. When out-of-control procedures are identified, the necessary corrective actions, also known as countermeasures, can be taken to ensure the product is safe. Self-inspection also aids regulatory agencies with ready access to documentation that food defense and food safety principles are practiced at all times.

Figure 1.4 Confidence is increased when there is a culture of commitment

What Should Be in a Food Defense SHARE Plan?

In the development of your food defense plan, you will need to use the SHARE principles as you evaluate your overall facility. As a reminder, the SHARE principles are strategy, hazards, accessibility, restriction, and enforcement. The key is to use your written food defense plan to compare and evaluate it against the actual systems in place, such as your processes, behaviors, and human resources.

The key is to use your written food defense plan to compare and evaluate it against the actual systems in place, such as your processes, behaviors, and human resources.

A written Food Defense SHARE plan should include the following steps:

(1) **Vulnerability Assessment.** It is critical to conduct a vulnerability assessment as your first step in each component of the SHARE plan. This is the process used to identify weaknesses and top physical security priorities in your operation.

(2) **Mitigation Strategies.** The second step is a description of corrective actions, or **countermeasures**, which are also known as mitigation strategies. Mitigation strategies are solutions for each area of weakness identified during the vulnerability assessment. An example of a mitigation strategy is this: Locking the back door of your operation at all times. Later on in this manual, we will discuss mitigation strategies in greater detail as it pertains to a retail foodservice operation.

Mitigation strategies are solutions for each area of weakness identified during the vulnerability assessment.

(3) **Traceability.** The third step is a description of a facility's procedures for tracing distribution. Traceability of your distribution requires having the ability to identify where your food products and/or ingredients were received from and where they are going when they leave your possession. In other words, this is one distribution level forward and one distribution level backward, as required by the **Bioterrorism Act of 2002 and the *Reportable Food Registry (RFR) for Industry*** of the **Food and Drug Administration** (FDA). In a retail or foodservice operation, your one level back could be the distribution center, wholesaler, supplier, vendor, or farmer. Your one level forward is where your food is going, which could also be another business. In a commissary kitchen, the one level back is the distribution center and the one level forward is your retail or foodservice operation. It is important to remember that traceability includes internal tracking of your food and ingredients.

(4) **Verification.** The fourth step is description of the verification procedures that must be completed annually as part of your HACCP and Food Defense SHARE plans.

Keep in mind, you cannot have the same person responsible for monitoring and also conducting verification reviews.

(5) Recordkeeping and Documentation. The last step is a description of the facility's food defense recordkeeping procedures, which is similar to HACCP documentation. Be sure to include a description of the facility's procedures for food recalls and whether those recalls are due to unintentional or intentional contamination. These procedures should also contain health surveillance protocols for monitoring and following up with sick employees and/or absenteeism.

HUMAN RESOURCE SYSTEMS

Here are some important questions to consider on how people factor into food defense procedures at a facility. If you answer *no* to any of these questions, we recommend you discuss this with the management in your organization or consult your attorney.

Does your company conduct employee **background checks** as a condition of employment? Does your company have procedures concerning cooperation during an investigation? Is food defense addressed in your employee handbook?

Does your company use **confidentiality and nondisclosure agreements** as part of your human resource protocol? Is food defense addressed in your job descriptions? Are job descriptions accurate and current? Has your human resource process been reviewed by an attorney concerning food defense? Can you provide records and documentation that food defense is a part of your human resource process? Is your organization able to support a qualified leader in food defense?

A qualified leader is one that has the power of authority and the duty to act.

It is critical to have a qualified leader responsible for food defense. A qualified leader is one that has the power of authority and the duty to act. This qualified leader may be added to your business as long as the appropriate resources are allocated. Some of these resources can be a challenge based on cost and complexity. The goal is to mitigate vulnerabilities within your control. No matter the situation, this cannot be ignored. The notion that "It won't happen to me (or to the organization)" has failed time and time again. It is critical to set the right food defense tone for your organization using a qualified leader, awareness, and communication techniques. This is the time to be proactive and assess your people and facility as if your life or livelihood depended upon it.

Technology, instant messaging, communication, and social media networks have changed the world in profound ways. As it relates to food terrorism, the need to disrupt and prevent attacks will hinge on your ability to adopt technology, instant messaging, communication, and social media networks from farm to fork. The days of flying under the radar screen and hoping that something will go unnoticed are over. For example, if an organization thinks the best method is "do not disturb" and follows a habit of "let sleeping dogs lie," then the organization could potentially destroy its brand or lives could be lost.

Every employee in your organization must be responsible for food defense and understand some basic physical security practices, including awareness of coworkers, customers, vendors, and the facility.

Food defense is the act of preventing the deliberate contamination of food through awareness programs, such as Food Defense SHARE. Every employee in your organization must be responsible for food defense and understand some basic physical security practices, including awareness of coworkers, customers, vendors, and the facility. Given below is an employee food defense standard operating procedure (SOP) form that your organization can use during orientation and training, or in your existing human resource system. The employees food defense SOP form strengthens your food defense culture. The following are the food defense awareness standard operating procedures (SOPs) provided for your use as a courtesy from *The HACCP Food Safety Training Manual* which outlines awareness action steps and procedures for employees, customers, vendors, and facility:

Employee Awareness SOP

- Be a responsible employee. Communicate any potential food defense issues to your manager.
- Be aware of your surroundings and pay close attention to customers and employees who are acting suspiciously.
- Limit the amount of personal items you bring into your work establishment.
- Be aware of who is working at a given time and where (in what area) they are supposed to be working.
- Periodically monitor the salad bar and food displays.
- Make sure labeled chemicals are in a designated storage area.
- Make sure you and your coworkers are following company guidelines. If you have any questions or feel as though company guidelines are not being followed, please ask your manager to assist you.
- Take all threats seriously, even if it is a coworker blowing off steam about your manager and what he or she wants to do to get back at your manager or your company; or if he or she is angry and wants to harm the manager, the customer, or the business.
- If the back door is supposed to be locked and secure, *make sure it is!*
- If you use a food product every day and it is supposed to be blue but today it is green, stop using the product and notify your manager.
- If you know an employee is no longer with your company and this person enters an "employees only" area, notify your manager immediately.
- Cooperate in all investigations.
- Do not talk to the media; refer all questions to your corporate office.
- If you are aware of a hoax, notify your manager immediately.

Customer Awareness SOP

- Be aware of any unattended bags or briefcases customers bring into your operation.
- If a customer walks into an "employee only" area of your operation, ask the customer politely if he or she needs help, then notify a member of management.

Vendor Awareness SOP

- Check the identification of any vendor or service person that enters restricted area of your operation and do not leave him or her unattended.
- Monitor all products received and look for any signs of tampering.
- When a vendor is making a delivery, never accept more items than what is listed on your invoice. If the vendor attempts to give you more items than what is listed, notify your manager.
- When receiving deliveries:

Step 1: Always ask for identification.
Step 2: Stay with the delivery person.
Step 3: Do not allow the person to roam freely throughout your operation.

Facility Awareness SOP

- Report all equipment, maintenance, and security issues to your manager.
- Document any equipment, maintenance, and security issues.
- Be aware of the inside and outside of your facility, including the dumpster area, and report anything out of the ordinary.

How Do You Create a Food Defense Culture That Is Conscientious, But Not Alarming?

The "If You See Something, Say Something" campaign was originally implemented by New York City's Metropolitan Transportation Authority and funded, in part, by the Department of Homeland Security (DHS). "If You See Something, Say Something" is a

"If You See Something, Say Something" is a simple and effective concept to engage the public and key employees to identify and report suspicious activity, and indicators of terrorism, crime, and other threats to the proper law enforcement authorities.

simple and effective concept to engage the public and key employees to identify and report suspicious activity, and indicators of terrorism, crime, and other threats to the proper law enforcement authorities. The DHS is committed to expanding the "If You See Something, Say Something" campaign nationally, with public education materials and outreach tools designed to help businesses, communities, and citizens who remain vigilant and play an active role. It is a simple way to reinforce that everyone in this world needs to be aware of their surroundings at all times. If you observe something out of the ordinary, you need to speak up. The purpose of this vigilance is to deter and stop domestic terrorists. Examples of potential concerns are an unattended bag or box or individuals behaving in a suspicious manner.

The ideal food defense culture is a conscientious working environment that does not cause panic and is not alarming.

The ideal food defense culture is a conscientious working environment that does not cause panic and is not alarming. In the hospitality industry, your organization's attitude by adopting "If you see something, say something" is best approached by the philosophy of "This is how we do business. The expectation is for you to remain calm and notify your manager immediately."

Another food defense concern is the proper use, handling, and storage of chemicals. Chemical awareness training should already be in place with the laws pertaining to employee safety. "Right to Know" is a law that requires employers to tell employees the hazards that they may be exposed to in the workplace using material safety data sheets (MSDS) that explain the details of the chemical and the potential hazards. This is simply completed by conducting training sessions and meetings. It is also necessary to document that it has been completed similar to food defense. In the government, the "Need to Know" is very important in protecting national security, just as the "Right to Know" is in protecting your employees.

Food Defense SHARE, uses monitoring as a method of managing, controlling, or observing an activity. Monitoring your operation to determine suspicious activity will help identify circumstances that may warrant an investigation or employee action. The use of observation, documentation, and records is necessary to determine if monitoring food defense activities is adequate throughout the flow of food.

- Observations may involve a range of methods such as informal interviews, direct observation, discussions, and analysis of documents.
- Documentation may involve methods such as document drafting, editing, formatting, submitting, reviewing, approving, distributing, and tracking.
- Records may include document verification activities such as vulnerability assessments, baseline physical security surveys, a written food defense plan, results of ongoing verification, and documentation from the reassessments.

Let's examine some examples of food defense audit protocols for human resources, managers, employees, and physical security.

Human Resource Food Defense Audit Protocol

Here is a sample item included in a human resource audit checklist:

Are employees receiving training in food defense, and is the training documented?

- Review your food defense plan to ensure that provisions are made to train employees in their food defense responsibilities.
- Verify employee training records, including names, dates, and instructors.
- Ask employees about their role in food safety and food defense.

Manager's Food Defense Audit Protocol

Here is a sample item included in a manager's audit checklist:

Figure 1.5 Managers of a restaurant support If You See Something Say Something.

Are all shipments inspected upon arrival?

- Review the food defense plan to ensure that it provides guidance to managers on inspecting shipments upon receiving.
- Review receiving protocols for examining shipments for discrepancies such as incomplete information on forms and how they are handled.
- Document when procedures are not being followed, such as how incomplete forms are mishandled.

Employee's Food Defense Audit Protocol

Here is a sample item included in an employee's audit checklist:

Is the food defense SOP orientation form completed for all employees?

- Review the *food defense SOP orientation form* of each employee.
- Review employee training records and program documents.
- Ask employees if food defense training was provided by management.

Physical Security Food Defense Audit Protocol

Here is a sample item included in a physical security audit checklist:

Are employees trained on entering or leaving a restricted area for storage?

- Review the food defense plan procedures that detail the frequency and thoroughness of storage inspections.
- Review inspection reports to determine if employees are completing the storage logs at the frequency set in the food defense plan.
- Observe employees entering the restricted areas of the facility and make note when procedures are not being followed.

Recognize the Importance of Food Defense Strategy

Culture is simply the way an organization conducts or does business and how the business can adjust and change business practices.

Share this prevention strategy with your most valuable resource: your people. The prevention of the deliberate contamination of food should be a significant part of your organization's culture as it pertains to strategy. Culture is simply the way an organization conducts or does business and how the business can adjust and change business practices. In a systematic human resource approach, a business traditionally makes use of recruiting, hiring, and training, accompanied by policies and procedures. Awareness should be reflected as part of your strategic plan for ensuring your people and foods are safe and secure.

It is essential to take everything that you have learned in this chapter with you and apply it to the flow of food. The "flow of food" approach provides us with insight into how foods flow from purchase, receive, store, prepare, cook, hold, cool, reheat, and serve at any given time. People are the key ingredient. That's why you need to have a strong strategy in your organization.

In the next chapter, we will focus on the second element in SHARE, which is the hazards in your operation. The hazards element of the plan will focus on preventing intentional contamination from biological and chemical agents at a workstation, as well as in your operation, by following the flow of food.

Strategy Study Questions

1. In a Food Defense SHARE plan, what are the five components?

   a. ______________________________

   b. ______________________________

   c. ______________________________

   d. ______________________________

   e. ______________________________

2. What is the difference between food defense, food safety, and HACCP?

   ____

3. What are the five steps to a written Food Defense SHARE plan?

a.

b.

c.

d.

e.

4. In food defense it is critical to have a conscientious leader and/or manager. What is a qualified food defense leader?

5. How does information sharing relate to food defense?

APPLY YOUR KNOWLEDGE
(Close Hold Assignment)

1. What are some barriers that you can break down to help your organization apply food defense?

2. How would you approach applying a reward and discipline process pertaining to food defense?

Congratulations! You have successfully completed *Strategy*, the first step in SHARE! Let's begin with Chapter 2, Hazards . . .

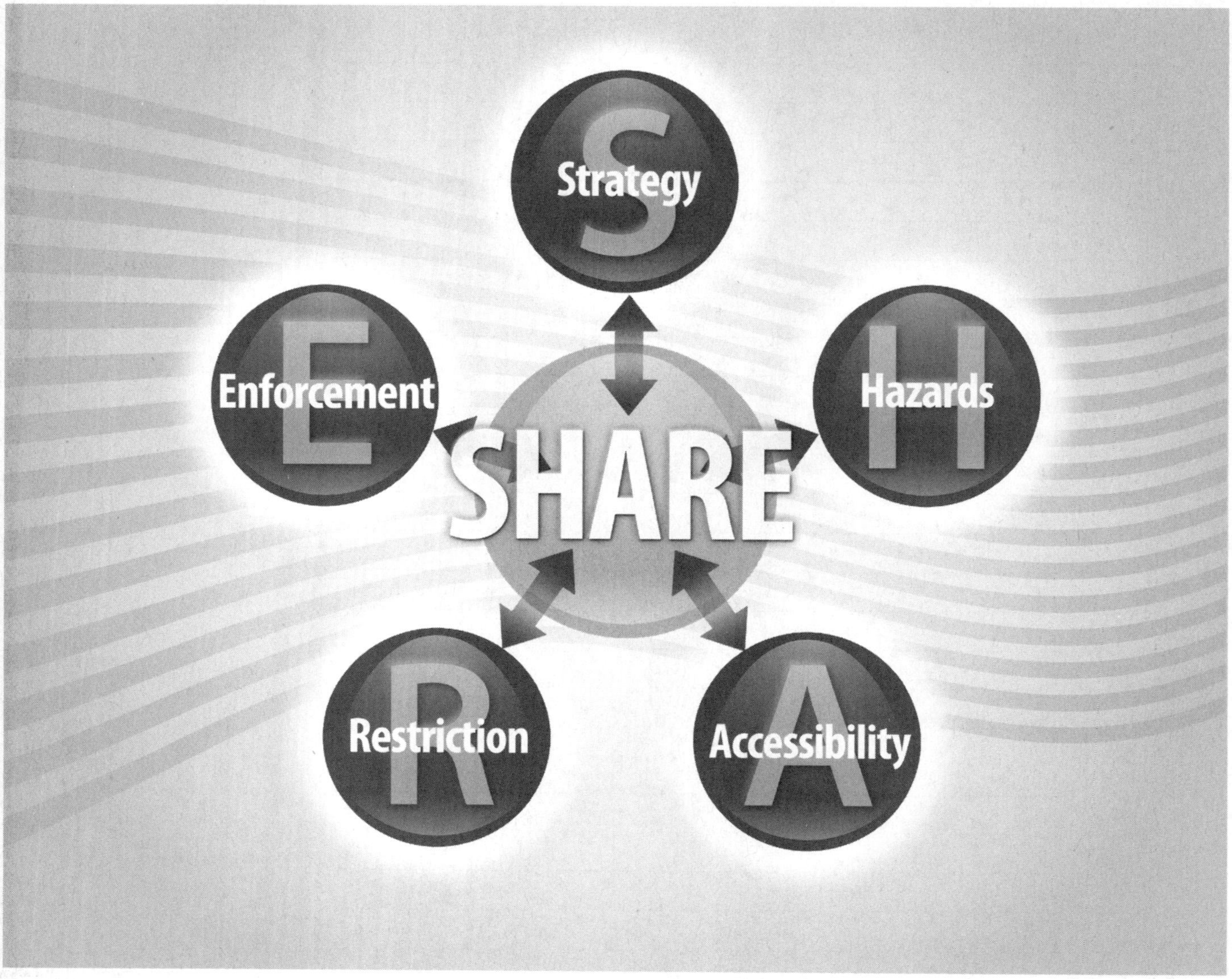

Chapter 2
Hazards

OVERVIEW

Protecting the food supply is not an option; it is nonnegotiable. It is essential to understand the potential negative impact caused by the introduction of hazards in the *flow of food* (i.e., product stream, or processing steps). The hazards may be introduced when you purchase, receive, store, prepare, cook, hold, cool, reheat, and serve food at any given point in your operation. Hazards are the second part in the Food Defense SHARE principles.

Biological and chemical hazards must be addressed, clearly understood, and managed as a normal course of business to improve food safety and food defense in your company.

Food safety hazards are typically biological, chemical, and physical. Biological and chemical hazards must be addressed, clearly understood, and managed as a normal course of business to improve food safety and food defense in your company. **Biological hazards** are viruses, bacteria, parasites, fungi, and mold; **chemical hazards** commonly used in the food industry include cleaning agents, sanitizers, and pesticides. Physical hazards are defined as foreign objects such as dirt, glass, fingernails, and metal fragments. Physical hazards are addressed from a HACCP food safety perspective and are *not* a component of a Food Defense SHARE plan.

Foreign and domestic terrorists have utilized **biological**, **chemical**, and **radiological hazards** as weapons. In the case of food defense, these terrorists are focused on causing mass destruction by the deliberate introduction of biological and chemical agents into food. These **agents** are poisonous or toxic substances (i.e., ricin or arsenic) that could harm the environment, animals, and humans, causing death, illness, panic, and mistrust. The deliberate or intentional introduction of an agent(s) can also increase the economic burden on the food and agriculture sector, industry, and government.

Radiological hazards are loose particles of radioactive material, such as plutonium and uranium. Radiological hazards are a priority for Food Defense SHARE; however, the easy of availability to be intentionally introduced into food is unlikely. Terrorists could cause radiological hazards by the deliberate introduction of radioactive material to plants, animals, humans, and/or objects.

Here is an example of an unintentional radiological contamination event: On March 11, 2011, a 9.0-magnitude earthquake followed by a tsunami hit Japan. Clearly, this tragedy was a natural disaster, which damaged six Japanese nuclear reactors. These nuclear reactors released radioactive material, resulting in a stop on imported Japanese food products (e.g., milk, milk products, fresh vegetables, and fruit). However, if an intentional contamination by a biological, chemical, or radiological agent occurs within the global food supply, then the outcome would be an epic disaster. This is another reason why it is critical to have an HACCP food safety plan and a Food Defense SHARE plan, as discussed in Chapter 1.

If vulnerabilities and hazards are appropriately identified, then you will be able to prevent, eliminate, and reduce risks.

A food defense SHARE plan must address vulnerabilities and hazards to help ensure a food defense culture for protecting food products from an intentional *adulteration and contamination* event. If vulnerabilities and hazards are appropriately identified, then you will be able to prevent, eliminate, and reduce risks. If vulnerabilities and hazards are not identified, then acts of terrorism and disaster can occur.

Carefully guarding and securing the information from **vulnerability assessments** and **close hold assignments** within the Food Defense SHARE program is recommended to protect your employees, customers, brand, and industry. Now, let's begin with the learning objectives of the second step in Food Defense SHARE for hazards.

LEARNING OBJECTIVES

In this chapter, you will learn to:

- Understand hazards as they relate to Food Defense SHARE.
- Recognize hazards and vulnerabilities in the flow of food in your foodservice operation.

- Explain the seven attributes of the CARVER+Shock method.
- Utilize the steps for conducting a CARVER+Shock analysis.
- Strengthen your food defense culture and infrastructure.
- Compare HACCP food safety and Food Defense SHARE principles.

EXERCISE: WHAT ARE YOUR VULNERABILITIES?

Let's assess whether your operation addresses hazards and vulnerabilities as they relate to Food Defense SHARE. The term **vulnerability** refers to areas of weakness in communication, infrastructure, physical building, equipment, and, more specifically, the potential exposure of foods to hazards. Each organization must *protect their assets and strengthen infrastructure* by analyzing potential hazards and vulnerabilities. Answer *yes* or *no* to the following questions about your operation. Place a check in the appropriate box. You will refer to this exercise later when completing your food defense plan.

Hazard Vulnerability Assessment	Yes	No
1. Does management recognize the potential biological hazards in your operation?	☐	☐
2. Does management recognize the potential chemical hazards in your operation?	☐	☐
3. Do you have preventative measures in place to protect human life?	☐	☐
4. Do you have preventative measures in place to ensure safety from hazards?	☐	☐
5. Do you have preventative measures in place to monitor hazards?	☐	☐
6. Do you have preventative measures in place to prevent hazards from entering your physical property?	☐	☐
7. Do you have corrective action steps to identify a system failure?	☐	☐
8. Does management recognize the vulnerabilities in your operation?	☐	☐
9. Are the hazards and vulnerabilities documented throughout your operation?	☐	☐
10. Are food defense procedures clear and practiced in your operation?	☐	☐
11. Are food defense procedures effectively communicated to all employees?	☐	☐
12. Has management or a third-party contractor conducted a vulnerability assessment?	☐	☐
13. Is there an established procedure in place for auditing food defense practices?	☐	☐
14. Can management verify if food defense practices are being followed?	☐	☐
15. Does management have open communication with employees?	☐	☐
16. Is your building physical structure properly maintained, in good repair and condition?	☐	☐
17. Is your equipment properly maintained, in good repair and condition?	☐	☐
18. Is the building and equipment clean and maintained in a sanitary condition?	☐	☐

- If you answered *yes* to *all* of the questions, congratulations—this is proof that your organization is focused on food defense.
- However, if you answered *no* to *any* of the questions, it is recommended that your organization develop or update their food defense plan.

FIGURE 2.1 The time is now to address food defense in your organization.

Understanding Vulnerability

Now that you have completed the assessment of your food defense hazards, it is important to understand the path going forward. Vulnerabilities are related to unintentional and intentional hazards in food. Food defense vulnerability is the ease or ability of an attacker to accomplish a desired outcome and refers to areas of weakness in communication, infrastructure, physical building, equipment, and, more specifically, the potential exposure of food to hazards. A vulnerability assessment (VA) is a tool used in food defense to simply analyze the entire operation for weaknesses. This process helps to identify (those) weaknesses and top physical security priorities in your operation. A vulnerability assessment should capture your entire process (i.e., flow of food) and assign a relative risk ranking score per workstation, or food handling area. This ranking also helps in the prioritization of resources to secure your operation. The expectation for every operation is that the person in charge, such as a conscientious leader and/or manager, will apply and conduct a vulnerability assessment for and throughout the flow of food.

A vulnerability assessment (VA) is a tool used in food defense to simply analyze the entire operation for weaknesses.

Let's review the flow of food.

PURCHASE
↓
RECEIVE
↓
STORE
↓
PREPARE
↓
COOK
↓
HOLD
↓
COOL
↓
REHEAT
↓
SERVE

Within a retail foodservice operation, the flow of food begins after the recipes are carefully selected and food is purchased from an approved reputable supplier. An approved reputable supplier must be in compliance with local, state, federal, and/or international laws. The retail foodservice operation then receives and stores the food properly. The food may then go through preparation, cooking, holding, cooling, and reheating prior to the last step in the flow of food, which is serving.

FIGURE 2.2 Food defense awareness is important especially when food is open or accessible to the public.

There are four major vulnerability assessment factors of concern in a Food Defense SHARE plan:

(1) *products with a short shelf life*,

(2) *uniformly mixed products*,

(3) *products with large batch sizes*,

and (4) *easily accessible parts of your operation*, meaning areas with a high degree of openness.

Determining the most vulnerable points in your infrastructure enables you to focus your limited resources on protecting your most vulnerable points, which will have some overlap with HACCP food safety. There are four major vulnerability assessment factors of concern in a Food Defense SHARE plan: (1) *products with a short shelf life*, (2) *uniformly mixed products*, (3) *products with large batch sizes*, and (4) *easily accessible parts of your operation*, meaning areas with a high degree of openness. These factors assist you to target mitigation strategies for the most vulnerable points in your operation.

Reality Check

Is Food Contamination a Credible Threat?

The Department of Homeland Security has said that al Qaeda, which operates out of the Arabian Peninsula, has developed terrorist tactics that include spreading such poisons as ricin and cyanide in salad bars and buffets at U.S. restaurants. Some U.S. officials have claimed that this Middle Eastern terrorist group not only is capable of but also has plans to poison our food supply. U.S. officials have responded by meeting with members of the hospitality industry to stress the possibility of a terrorist threat to our food supply. They have also outlined the best practices for keeping the food supply safe in hotel and restaurant channels.

The Department of Homeland Security's official comment came in response to a CBS report. "We are not going to comment on reports of specific terrorist planning. However, the counterterrorism and homeland security communities have engaged in extensive efforts for many years to guard against all types of terrorist attacks, including unconventional attacks using chemical, biological, radiological, and nuclear materials," spokesman Sean Smith said in a prepared statement.

The focus is on the need for the hospitality industry to be aware of hazards and accept the fact this is a credible threat to the foodservice industry (business).

(continued)

The latest hazards of choice happen to be ricin and cyanide intentionally introduced into the global food supply by an attacker. The Centers for Disease Control and Prevention (CDC) published the following scientific information on their Web site about ricin and cyanide, respectively:

> Ricin is a poison found naturally in castor beans. Ricin can be made from the waste material left over from processing castor beans and then purified, and treated to form a powder that can be inhaled although this is a very technically difficult and complicated process. Most ricin poisonings have occurred when the ricin was injected or when the person swallowed the ricin. Symptoms of ricin poisoning are most likely to occur within 4 to 12 hours if the ricin was inhaled or swallowed. Ricin poisoning is not contagious. It cannot be spread from person to person through casual contact. It would take a deliberate act to make ricin and use it to poison people. Accidental exposure to ricin is highly unlikely, except through the ingestion of castor beans. If made into a partially purified material or refined into a terrorist or warfare agent, ricin could be used to expose people through the air, food, or water.
>
> Cyanide is a rapidly acting, potentially deadly chemical that can exist in various forms. Cyanide sometimes is described as having a "bitter almond" smell, but it does not always give off an odor, and not everyone can detect this odor. You could be exposed to cyanide by breathing air, drinking water, eating food, or touching soil that contains cyanide. Cyanide enters water, soil, or air as a result of both natural processes and industrial activities. In air, cyanide is present mainly as gaseous hydrogen cyanide. Smoking cigarettes is probably one of the major sources of cyanide exposure for people who do not work in cyanide-related industries. Cyanide is more harmful to the heart and brain than to other organs because the heart and brain use a lot of oxygen.

If we fail to identify potential hazards, conduct vulnerability assessments, create an effective food defense culture, plan, and institute self-inspections and third-party audits, then our ability to deter an attack could be inhibited, which could result in mass casualties. Becoming more vigilant every day and incorporating Food Defense SHARE principles is essential for all organizations from farm to fork. Let's look at some key terms to help understand hazards and your role.

Key Terms Match Game

In the space provided, fill in the blank with the appropriate word that best fits the sentence as it relates to hazards.

Traceability
Accessibility
Recuperability
Vulnerability Assessment
Effect
Recognizability
Node
Shock
Biological Hazards
Chemical Hazards

1. ______________________________
is a consequence that follows an event, such as the direct loss in production.

2. ______________________________
is the ability of your operation to recover to normal production levels after an event.

3. ______________________________
is the ability to identify a target with little or no additional training.

4. ______________________________
is the ability to track food from farm to fork.

5. ______________________________
is the ability to physically get in and get out of a critical work area without being detected.

6. ______________________________
are dangerous substances that are in the category of viruses, bacteria, parasite, fungi, and mold.

7. ______________________________
are dangerous substances in the category of cleaning agents, sanitizers, and pesticides.

8. ______________________________
is a tool used in food defense to simply analyze the entire operation for weaknesses.

9. ______________________________
is the accessible parts of an operation.

10. ______________________________
is a combination of health, economics, and psychological impacts.

LEADERSHIP, SAFETY, AND SECURITY PROTOCOLS

What's Next? Where to Get Started with Hazards?

Let's begin with understanding *H*azard *A*nalysis and *C*ritical *C*ontrol *P*oint (HACCP), global food supply, food defense (as it relates to HACCP food safety), and the HACCP Star.

HACCP is a written food safety system to enable the selling and serving of safe food. HACCP prevents disasters, such as foodborne illness outbreaks, from occurring in the flow of food. The HACCP food safety philosophy simply states that biological, chemical, or physical hazards, at certain points in the flow of food, can be prevented, eliminated, or reduced to safe levels.

Food defense is a **prerequisite program** that is required for all HACCP food safety plans. A prerequisite program is a standard operating procedure that addresses basic operational conditions in an establishment.

Food Defense and the Global Food Supply

Why should you care about the global food supply? Today, foods are transported around the world more than ever before and there is unlimited seasonality of available food products as compared to years ago when there were distinct absences of specialty food items at certain times of the year. As a result of globalization, more people and machines are handling food products and there is a greater exposure opportunity for contamination to occur. This greater opportunity to spread adulterated products can be made even worse by intentional contamination and could result in mass causalities from foodborne illness, disease, and even death throughout the global food supply.

Every person is at risk of contracting a foodborne illness. Food can become contaminated either unintentionally or deliberately: "Increasing industrialization and urban living has meant that the food chain has become longer and more complex, increasing opportunities for contamination. It also means that more people are likely to be affected by a single breakdown in food hygiene" (WHO). Knowing the impact of foodborne illness helps to *rank risk*, *change policy*, *develop procedures*, and *prioritize practices.*

Can you guess the public health and economic impact for the following statements?

- In the United States, there are ________________________ cases of foodborne illnesses, resulting in ________________________ hospitalizations, and ________________________ deaths are estimated to occur each year.
- Foodborne illnesses in the United States are estimated to cost ________________________ in medical costs and lost productivity.
- Foodborne illnesses account for about 1 of every ________________________ U.S. hospitalizations and 1 of every ________________________ U.S. deaths.
- The FDA estimates percent ________________________ of foodborne illness victims develop secondary long-term medical issues that result in over ________________________ lingering health problems per year.

Every participant in the global food supply should be knowledgeable of the vulnerabilities that exist throughout the flow of food. It's our priority in this chapter to make it clear how the same approach is used in both HACCP food safety and Food Defense SHARE. This is critical for not only *moral*, *ethical*, *social*, and *economic* reasons but also *legal* reasons. The institutions that sell food have strict liability in these matters. Strict liability means that there is case law that holds the ownership and management of the operation liable for any death, injury, or damage caused by the adulterated food products. An example of strict liability in a different area of concern is with a dog bite. In most states, if you own a dog and your dog bites someone, you are liable. Whether the dog has a tendency to bite or not, you are still responsible if your dog bites.

Strict liability means that there is case law that holds the ownership and management of the operation liable for any death, injury, or damage caused by the adulterated food products. An example of strict liability in a different area of concern is with a dog bite. In most states, if you own a dog and your dog bites someone, you are liable.

Vulnerability in our food supply in one location can have multiple public health and economic impacts. An example of this occurred when one farm was identified as the source of contaminated fresh spinach. This single farm source of contaminated spinach resulted in having all fresh spinach sales stopped within the United States for a period of time. It disrupted all farms and food industry businesses that grew, handled, or sold fresh spinach.

Food defense events or foodborne illness outbreaks impact owners, corporate officers, and management officials, as they are held accountable whether the cause is known or unknown. Each industry also has its own varied public health and economic impacts from foodborne illnesses.

Some industry-specific situations provide examples of potential exposure due to the nature of their business or the impact of the services they offer.

Tourist Population: Airports, Cruise Ships, Hotels, and Casinos

AIRPORTS

- Most foodborne illnesses are not traced to their original source. They are even harder to trace in airports because customers are in transit to locations throughout the country and the world.
- Travelers are eating more in airports than in the past because of the recommended arrival time of a 2-hour minimum before their flight, and airlines are serving less food on board.

FIGURE 2.3 Airports are focused on food defense and passenger safety throughout the world.

- There are 19,820 airports in the United States.
- In 2005, 42,402,653 passengers traveled through Atlanta's Hartsfield-Jackson International Airport (United States' busiest airport).
- In 2009, a total number of 764,807,220 passengers traveled on commercial flights in the United States.

CRUISE SHIPS

- Outbreaks are of particular public health importance because of the confined spaces on cruise ships and because people eat and drink from a single food and water supply.
- More than 25 percent of cruise ship foodborne illnesses have been caused by salmonella.
- Cruise ships are in an ideal location to investigate foodborne illness outbreaks because they are self-contained.
- There are strict reporting requirements for foodborne illnesses on cruise ships, which leads to outbreaks getting a lot of attention in the media.
- A Norovirus outbreak on a cruise ship led to the shortening of their schedule so that the ship could go through a 72-hour cleaning.

FIGURE 2.4 Cruise ships around the world are taking initiatives in food safety and defense to protect their passengers.

HOTELS

- Even as far back as 1988, there were over 164 people infected by a foodborne outbreak. Hotels are of particular importance because many different categories of people such as the staff, overnight hotel guests, and people attending functions held in the hotel have the potential of getting sick.

FIGURE 2.5 Customers expect the hospitality industry to be leaders in food defense.

CASINOS

- Casinos are much like hotels because of the use of meeting and sleeping rooms and restaurants used by people who will be traveling after exposure to foodborne illness agents.

FIGURE 2.6 The use of technology has redefined ways to monitor and enforce food defense and physical security.

Colleges/Universities

COLLEGES

- Students should practice proper food safety. They should be especially careful because it is often the first time they are preparing food for themselves.
- The number of universities in the world is 8,571 in 201 countries as of October 2010.
- The national average of people willing to eat pink, or undercooked, hamburger is 19.7 percent; in 2002, 43.5 percent of college students engaged in this risky behavior.

FIGURE 2.7 Colleges and universities have elevated activities to protect the student body.

At-Risk Populations: Hospitals, Schools, and Senior Care Facilities

HOSPITALS

- According to the U.S. census, hospitals employ 5.1 million people nationwide and on any given day there are 539,000 hospital inpatients. These are all potential consumers of hospital food that could be affected by a foodborne illness.

FIGURE 2.8 Advocates of highly susceptible populations use food defense as a normal business practice.

SCHOOLS

- Each day more than 31 million children get their lunches through the National School Lunch Program.
- The National School Lunch Program cost $9.8 billion in FY 2009. By comparison, the lunch program's total cost in 1947 was $70 million; in 1950, $119.7 million; in 1960, $225.8 million; in 1970, $565.5 million; in 1980, $3.2 billion; in 1990, $3.7 billion; and in 2000, $6.1 billion.
- There were 97,382 public schools in the United States in 2005.

FIGURE 2.9 The National School Lunch Program is focused on protecting schoolchildren from both unintentional and intentional contamination while strengthening them nutritionally.

SENIOR CARE FACILITIES

- More than 1.8 million people live in nursing homes.
- Tens of thousands of older adults live in adult foster care facilities and group homes, while over 625,000 people live in assisted care facilities.
- According to U.S. Census Bureau projections, the older population (55 years and older) will double in 2030 to 72 million.

FIGURE 2.10 At all stages of care, food defense is critical.

Quick Service, Casual Dining, Fine Dining, Food Service Operations: Restaurants

RESTAURANTS

- According to the National Restaurant Association, in 2011 the Restaurant Industry sales were $604 billion dollars in the United States with 960,000 locations and 12.8 million employees. The overall economic impact of the restaurant industry is $1.7 trillion dollars.

Retail Food Outlets: Convenience Stores, Grocery Stores, and Supermarkets

CONVENIENCE STORES

- The U.S. convenience store industry accounts for more than $624 billion in sales for 144,541 convenience stores across the United States.
- The percentage of one-store operators, true "mom-and-pop" stores, continues to climb. The percentage of one-store operators topped 50 percent for the first time in 2001. Today, that percentage is 62.3 percent. The states with the highest percentage of one-store operations are Washington (78 percent), Georgia (74 percent), Louisiana (71 percent), Mississippi (71 percent), and New Jersey (71 percent). In Washington DC, 78 percent of convenience stores are one-store operations.
- Retail food types, also known as channel count comparison (as of December 31, 2009), are as follows:
    - Convenience stores: 144,541
    - Supermarkets: 35,612
    - Drug stores: 37,654
    - Mass merchandiser/dollar stores: 27,247
    - Wholesale clubs: 1,207

FIGURE 2.11 Convenience stores are expanding technology and training to address food defense for the protection of their customers, products, and brand.

GROCERY STORES

- In grocery stores, consumers pick their own food for purchase and preparation. This is opposed to eating in restaurants, which serve products that they have chosen. It would be beneficial to the general public to have knowledge of food safety when they select and purchase food from grocery stores and restaurants.

Government: Commissaries and Military Restaurants

- Defense Commissary Agency's Director of Health and Safety says that military and civilian health specialists work to assure that the commissary's supplies are as safe as possible. They only buy from approved sources.
- Customers are encouraged to follow food safety practices at home to ensure their health.

FIGURE 2.12 Grocery stores are being designed with food defense in mind.

FIGURE 2.13 Food defense standard operating procedures must protect the ones that protect us the most. Thank you for serving.

CARVER+Shock Methodology

CARVER+Shock is a methodology used by the food and agriculture (FA) sector to assess key issues or weaknesses related to accessibility and vulnerability in a food-service operation. This tool captures and ranks a foodservice production system, or steps throughout the flow of food in an operation. This method has helped the foodservice industry understand critical issues within the flow of food by assessing the level of *susceptibility* within a food production or processing system. It allows for an assumption of the *worst-case scenario* (i.e., *an inside attacker*) by identifying the most attractive *nodes* (nodes being the accessible parts throughout the flow of food), where there is direct human contact with food. For example, the person who may cause the most damage to public health, including the associated economic impact, by an intentional adulteration may have intimate knowledge of your operation. As a result, the protection of your operation will depend on your ability to deter the inside attacker.

This method has helped the foodservice industry understand critical issues within the flow of food by assessing the level of *susceptibility* within a food production or processing system.

CARVER+Shock stands for the following *seven attributes* used to evaluate the attractiveness of a node (based on worst-case scenarios). CARVER is an acronym used to evaluate the attractiveness of a target for attack:

Criticality—measure of public health and economic impacts of an attack

Accessibility—ability to physically access and egress from target

Recuperability—ability of system to recover from an attack

Vulnerability—ease of accomplishing attack

Effect—amount of direct loss from an attack as measured by loss in production

Recognizability—ease of identifying target

+ *SHOCK*—the combined health, economic, and psychological impacts of an attack

Steps for conducting a CARVER+Shock analysis

Step 1 **Assemble Experts.** A team of subject matter experts is essential to provide technical knowledge and accuracy for the assessment. The expert team should consist of individuals who are competent in the processing and production at your facility. In large operations, this team is specific to the food process or product being evaluated, food and nutritional sciences, *toxicology, epidemiology, microbiology*, and *risk assessment*. The team will apply the CARVER+Shock criteria to create the actual assessment.

Step 2 **Define the Flow of Food.** The analysis begins by describing the steps in the process under consideration for the assessment. A visual diagram of the flow of food such as a product flow chart of the system is required for an assessment. The nodes of the highest *vulnerabilities* must have *countermeasures*. Nodes are the possible individual pieces of equipment or accessible parts of the process. The scoring is repeated for each node or piece of equipment within the flow of food.

Step 3 **Establish the *Parameters*/Assumptions.** Before scoring the nodes, you should establish the process flow diagram specifically for each piece of equipment, as well as the assumptions you wish to use. Simply list all pieces of equipment or workstations. This is based on the *worst-case scenario*; the determining factors should include what you are protecting and whom you are trying to protect. For example, factors such as an aggressor using biological or chemical hazards are frequently used for worst-case scenarios.

Step 4 **Assign Scores.** The nodes should be assessed for each of the seven CARVER+Shock attributes, and an overall score for that node should be obtained. The nodes with the higher overall scores are those that are potentially the most vulnerable and attractive nodes. The *final score* is calculated and captured to reflect the assumptions. In evaluating and scoring the attractiveness of a node, the ranking is based on a scale from (1) one to (10) ten. Conditions that are associated with *lower attractiveness* or *lower vulnerability* are assigned lower values (e.g., (1) one or (2) two), whereas conditions associated with *higher attractiveness* as a node or higher vulnerability are assigned higher values (e.g., (9) nine or (10) ten). The CARVER+Shock software tool is available at www.FDA.gov/FoodDefense for conducting a vulnerability assessment (VA) and determining the most vulnerable points in your food service business, which can help you to protect your most vulnerable nodes and focus your limited resources on them.

Step 5 **Discuss Lessons Learned.** Once the scoring of nodes in the system has been completed, a plan should be developed to consider countermeasures to minimize the attractiveness of the highest nodes. For example, if the node is evaluated as a (9) nine or (10) ten, such as an unsupervised high-volume salad bar, then countermeasures or mitigation strategies must be taken. These practical mitigation strategies are assigning an employee to maintain the salad bar, observe guest behaviors, and watch for suspicious or unusual activity, as well as, food product irregularities. Pf

Step 6 **Mitigation Strategies.** Countermeasures might include enhancements to physical security, personnel security, and operational security to help minimize access to the product or process. The Food Defense Mitigation Strategies Database is available at www.FDA.gov/FoodDefense for correcting vulnerabilities that were identified.

Step 7 **Best Practices.** You should work with your experts to drill down throughout your organization from top to bottom and beyond to identify needed consumer protection best practices. Also include collaborative industry partnerships and supporting federal, state, and local law enforcement efforts to ensure proper prosecution. This will help facilitate success in bringing any "bad guys" to justice. Best practices make certain that you are able to prosecute the domestic or international terrorist to the highest extent of the law. The best applied science for all retail foodservice operations is based onfollowing the 2009 FDA Model Food Code section 3-306.13: Consumer Self-Service Operations (C) CONSUMER self-service operations such as buffets and salad bars shall be monitored by FOOD EMPLOYEES trained in safe operating procedures.

CARVER+Shock Weaknesses

Many people complain that the CARVER+Shock method doesn't fit their foodservice production system. For this reason, we are redefining CARVER methodology to fit your organization. You will gain an appreciation of the CARVER+Shock software tool (available online) and how to use it to evaluate your operation. In addition, we will show you step by step how to discover the weaknesses in your foodservice and/or production system and how to improve food defense at each critical or most vulnerable workstation(s) in your operation. We have taken the CARVER+Shock methodology and assigned terms that provide you with an easy way of learning these concepts. Lastly, we have designed the course to support full comprehension of concepts based on your needs for conducting a vulnerability assessment.

What Is the HACCP Star?

The **HACCP Star** is a visual representation of the five points required for a successful HACCP food safety plan. Your HACCP food safety plan will fail if all five points are *not* executed properly. It is easy to picture and remember the HACCP principles when you use a five-point star. Following are the components of the HACCP Star:

Star Point 1—Food Safety Prerequisite Programs:

Master food safety

Star Point 2—Food Defense:

Apply food defense

Star Point 3—Create the HACCP Plan:

HACCP Principle 1: conducting a hazard analysis

HACCP Principle 2: determining critical control points

Star Point 4—Work the Plan:

HACCP Principle 3: establish critical limits

HACCP Principle 4: monitoring

HACCP Principle 5: taking corrective actions

Star Point 5—Checks and Balances:

HACCP Principle 6: verification

HACCP Principle 7: recordkeeping and documentation

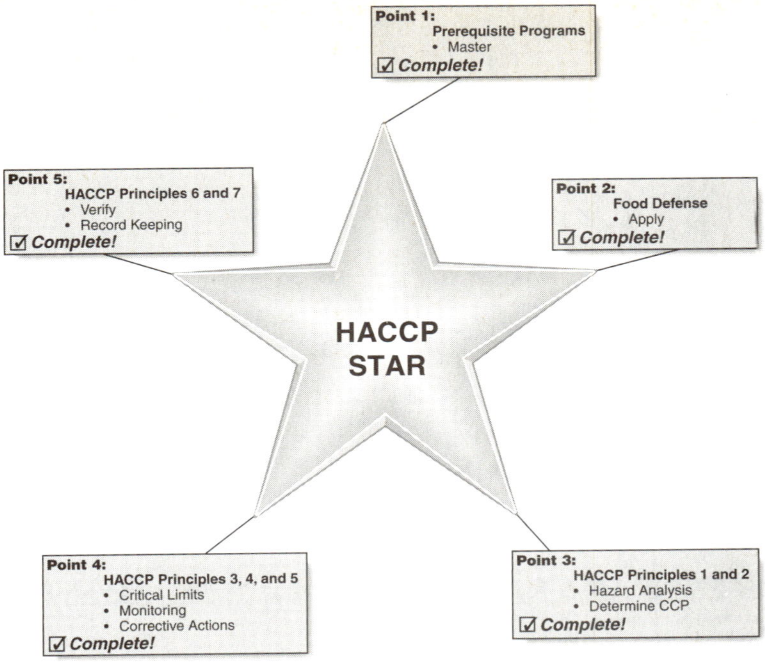

FIGURE 2.14 HACCP Star.

FOOD DEFENSE AND HACCP PRINCIPLE 1: CONDUCT A HAZARD ANALYSIS

A *hazard* in terms of food safety is characterized as any biological, chemical, radiological, or physical agent that may cause a food to be unsafe for human consumption. With food safety, a hazard is something that is reasonably likely to occur. Food defense goes further to consider worst-case scenarios and determine where a hazard is unreasonably likely to occur. This would require you to establish preventive controls and to be more proactive and decrease a reasonable possibility that the hazard will occur in the type of food being manufactured, packed, transported, prepared, or held by your company. The common goal that links together food defense and food safety is to deter, prevent, reduce, and eliminate hazards posed throughout the flow of food from farm to fork.

The common goal that links together food defense and food safety is to deter, prevent, reduce, and eliminate hazards posed throughout the flow of food from farm to fork.

A vulnerability assessment should help to identify the most important components of a functional food defense plan. It should also address the process on how to isolate, segregate, and decontaminate your foodservice system, or materials. For example, the identification of specific lot numbers is a critical issue to be considered in your recall and traceability sections of a functional food defense plan. The term *lot* or batch in food production is designated to quantities of food produced during a specific period of time. A lot can also refer to a supply of materials produced and used during a specific period of time. Lot numbers provide a specific code for traceability. *Traceability* and effective decontamination procedures for your facility are essential for business continuity, public health, incident management, and control of any event, or emergency.

Traceability and effective decontamination of your facility is essential for business continuity, public health incident management, and control of any event, both as part of your Food Defense SHARE plan and in the event of emergencies.

The CARVER+Shock methodology is similar to conducting a hazard analysis in HACCP food safety. The CARVER+Shock method is designated as a proactive, targeting, prioritization tool. This tool can be used to assess the vulnerabilities to an attack within a system or infrastructure. It allows you to think like an attacker by identifying the most attractive targets such as nodes or key workstations based on the likeliness or worst-case scenario.

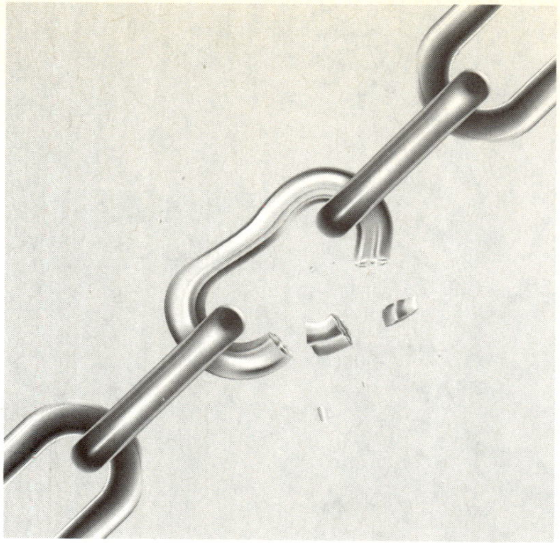

FIGURE 2.15 Food defense addresses the weakest links in your operation.

FOOD DEFENSE AND HACCP PRINCIPLE 2: DETERMINE CRITICAL CONTROL POINT

The flow of food is organized with each operational step as a control point or a critical control point based on the CARVER+Shock analysis.

Food defense critical control points (pieces of equipment, points, or nodes) are outlined in this book according to the flow of food for retail foodservice operations using food defense mitigation, prevention, response, and audits. The flow of food is organized with each operational step as a control point or a critical control point based on the CARVER+Shock analysis.

Keep in mind that although HACCP food safety and food defense have common factors, every operation is different, and critical control points will vary from one operation to another. In food defense, controls can be applied to prevent, eliminate, or reduce the hazard based on a vulnerability assessment.

FOOD DEFENSE AND HACCP PRINCIPLE 3: ESTABLISH CRITICAL LIMITS

In HACCP, establishing critical limits requires specific scientific measurements. An example of an HACCP food safety critical limit is cooking poultry to an internal minimum temperature of 165°F (73.88°C) for 15 seconds, the minimum temperature and time, respectively, required to destroy enough pathogens to ensure safety when eating the chicken.

Critical limits in food defense are attainable and realistic, and they must be met for each preventive measure.

In Food Defense SHARE, a critical limit is having two people always present to maintain dual control of your facility, better known as a buddy system. Critical limits in food defense are attainable and realistic, and they must be met for each preventive measure. If the critical limit is not met, a corrective action must be performed.

FOOD DEFENSE AND HACCP PRINCIPLE 4: MONITORING

In HACCP food safety, monitoring is best defined as conducting a planned sequence of observations or measurements to assess whether a process, point, or procedure is under control. Monitoring results in producing an accurate record for use in the next three HACCP principles of corrective action, verification, and recordkeeping.

The processes, procedures, and frequencies for monitoring the critical limit must identify any circumstances in which the critical limit preventive controls are not fully implemented. Monitoring processes and procedures is an ongoing activity performed in a continuous or noncontinuous manner and is always a work in progress. Continuous monitoring is preferred because it is constant monitoring of a critical control point or a specific node. This is done with built-in measuring equipment that records, for example, time and temperatures. Computerized equipment systems are an example of continuous monitoring. Monitoring is critical in both a food safety and food defense plan.

Monitoring processes and procedures is an ongoing activity performed in a continuous or noncontinuous manner and is always a work in progress.

Noncontinuous monitoring, also known as intermittent monitoring, occurs at scheduled intervals or on a per-batch basis and is performed often enough to ensure critical limits are met. This type of monitoring is primarily what the majority of food service operations use. An example of intermittent monitoring is using a properly calibrated, cleaned, and sanitized thermometer to measure the temperature of food products every 2 hours.

Monitoring in food defense requires active surveillance and may offer peer-to-peer incentives that can lead to apprehension of an attacker. In food defense, a manager monitors four basic factors. To do this, you can use technology and nontechnical tools such as simple visual observation. The four basic factors to consider are (1) products with a short shelf life, (2) uniformly mixed products, (3) products with large batch sizes, and (4) ease of access, focusing on areas with a high degree of openness. Technology is an amazing tool for food defense; video cameras, remote live feed, scanners, locks, and ID badges are important to deter potential food defense events. Retail foodservice operators have some catching up to do to improve monitoring of food defense activities through technology and surveillance. From a food defense and HACCP food safety perspective, many manufacturers and processors are pioneering the industry with innovative technologies, such as tattooing or branding of food with their logo, as in the stamp used on each pasteurized egg. Given below is the symbol used to identify eggs that were pasteurized and an alternative for this symbol is the actual word "pasteurized" on each egg.

FIGURE 2.16 This is a known symbol for egg pasteurization.

The reason why this branding of food is critical is that each individual egg has a potential biological hazard called salmonella that is found inside the egg. The process involved in the pasteurization of eggs, such as the "Pasteurized shell eggs" logo, means that these eggs, still in their shells, having been heat treated to destroy *Salmonella enteritidis* to the FDA standard of a 5-log reduction, are thus exempt from status as a potentially hazardous food because no viable *Salmonellae* exist.

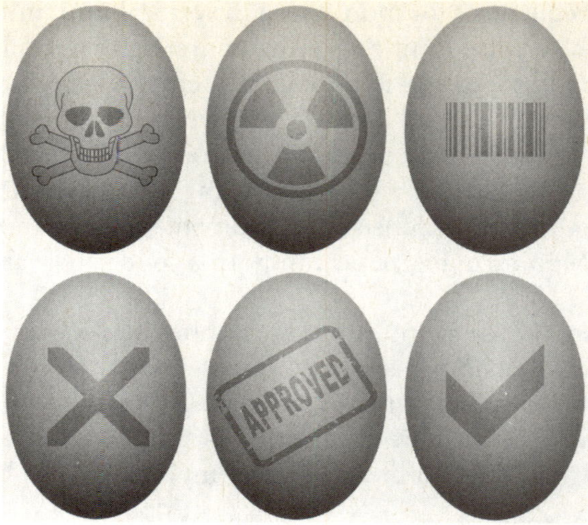

FIGURE 2.17 You must be vigilant at all times.

FOOD DEFENSE AND HACCP PRINCIPLE 5: TAKING CORRECTIVE ACTIONS

The corrective actions are predetermined steps that you automatically take if the critical limits are not being met.

The corrective actions are predetermined steps that you automatically take if the critical limits are not being met. This is Principle 5 of the HACCP food safety process. The five steps for corrective action that must occur for food defense are overlapping principles of HACCP food safety corrective actions.

The five steps listed below are necessary for an HACCP food safety corrective action:

- HACCP Step 1: To establish the exact cause of the deficiency.
- HACCP Step 2: To determine who is to correct the problem.
- HACCP Step 3: To correct the problem.
- HACCP Step 4: To decide what to do with the product.
- HACCP Step 5: To record the corrective action taken.

On the other hand, the following five Food Defense SHARE tasks are necessary for Food Defense SHARE corrective actions.

The Food Defense SHARE corrective action steps are as follows:

- **Food Defense Step 1**: To conduct a vulnerability assessment based on the degree of openness throughout the flow of food for each workspace/station/area. Then, determine the exact worst-case scenario for a breach in any physical security protocols that may warrant an investigation by law enforcement.
- **Food Defense Step 2**: To develop a written, functional food defense plan based on results obtained by performing a vulnerability assessment.
- **Food Defense Step 3**: To test or run drills of the written plan by monitoring periodically the effectiveness of protective measures and controls, countermeasures, mitigation strategies, and/or enhanced physical security measures to deter an inside attacker.
- **Food Defense Step 4**: To ensure the Food Defense SHARE plan is implemented when effective measures are identified. For example, you need to determine who will contact upper management and/or law enforcement.
- **Food Defense Step 5**: To maintain an executable, functional, and relevant Food Defense SHARE plan that will ensure that the protective measures, countermeasures, or mitigation strategies are reviewed and annually updated.

The greatest difference in HACCP food safety and Food Defense SHARE is the corrective action step to decide *what to do with the adulterated product.* In HACCP food safety, typical corrective actions are discarding the food, continuing to cook, reheating, and rejecting the food products. In food defense, corrective actions are different because the intentionally contaminated food may be needed as evidence for law enforcement, which will be explained further in Chapter 5. This is a huge mind-set change for the hospitality industry. In the hospitality industry, the culture dictates fixing the problem immediately and moving on to the next meal period or rush of business. In food defense, a manager needs to have a plan in place that is based on assumptions or worst-case scenarios should there be foul play. Then the food must be held and isolated to prevent additional contamination. Lastly, if a food defense event were to occur, a retail foodservice operation would *never* continue to cook or reheat the food products. The corrective action of rework and handling of food should be taken seriously. The nature of reworking is dangerous because the food is being handled again. In the event that food terrorism is suspected as part of your Food Defense SHARE procedures, the use of rework is not an option. Why? There is the potential, then, for an attacker to add a biological or chemical agent and adulterate the food that is being reworked.

The greatest difference in HACCP food safety and Food Defense SHARE is the corrective action step to decide *what to do with the adulterated product.*

Typically, rework of food means the food has quality or food safety concerns. Reworking food requires using clean, unadulterated food that is reconditioned so that it can be suitable for consumption. The challenging food safety decision associated with the reworking of food is whether or not to reuse the product or discard the food. This practice should be monitored and controlled in terms of your food reworking standard operating procedures and be clearly specific to food safety concerns only.

In food safety, taking a *corrective action for sanitation controls* should be described in the HACCP food safety plan and needs to include specific steps to be taken to eliminate the nonconformance or the failure to meet a specific requirement or critical limit. Corrective actions ensure that, if the preventive controls are not appropriately implemented, (1) no product enters commerce that is either harmful or adulterated, (2) the root cause is determined, and (3) finally, the failure is corrected.

Food defense corrective action steps to be taken shall include (1) segregating and holding the affected product, at least until a review is performed to determine the acceptability of the affected product for distribution; (2) taking actions necessary to ensure that intentional contaminated product never enters commerce; (3) identifying the root cause of the failure and correcting it; and (4) reducing the likelihood of recurrence through a breach in security protocols.

FOOD DEFENSE AND HACCP PRINCIPLE 6: VERIFICATION

Verification for food defense and HACCP food safety are very similar. Each one of these management systems must have a validation and verification process to make sure the food production system is working as intended. Whether it is HACCP food safety or food defense, the retail foodservice leadership must be confident that there are no gaps pertaining to employee, contractor, and/or customer.

Whether it is HACCP food safety or food defense, the retail foodservice leadership must be confident that there are no gaps pertaining to employee, contractor, and/or customer.

Ongoing verification should include (1) a review of any consumer complaints; (2) the calibration of process monitoring instruments and, when applicable, verification tools (questionnaires); (3) performance of periodic finished product or in-process testing to prove reduction of the hazard to an acceptable level; and (4) microbial testing for environmental pathogens. Lastly, the on-site verification audits are conducted by on-site auditors or approved third-party auditors/suppliers, and they meet the written product safety and defense specifications.

Reassessment of the HACCP food safety and Food Defense SHARE plans must include hazard analysis whenever any changes in the process or product occur. These plans should be revised immediately whenever new information is revealed, or an apparent need occurs.

These plans should be revised immediately whenever new information is revealed, or an apparent need occurs.

Food defense practices as an integral part of doing business will increase the need to conduct third-party audits, inspections, food defense surveys, and testing. *Verification* of a food defense plan ensures that the food defense plan is properly implemented. As *verification* procedures are developed for food defense, they will also be important to ascertain and determine whether a mitigation step is or has been operating as intended.

Validation means a preventive control that, when properly implemented, is capable of providing mitigation strategies geared to a specified outcome. Validation of the food safety plan should control the hazards and identify mitigation strategies that are needed prior to implementation, as well as those strategies that are needed when-ever a reassessment is required for the HACCP food safety and Food Defense SHARE plans.

FOOD DEFENSE AND HACCP PRINCIPLE 7: RECORDKEEPING AND DOCUMENTATION

All records must be accurate and legible.

It is critical when documenting HACCP food safety to get enough information to ensure your plan is effective; the same is true for food defense. All records must be accurate and legible. As the manager or supervisor, it is your responsibility to ensure correct and accurate recordkeeping. The employees on your team will follow your example. For instance, never scribble or use correction tape or liquid because it may look like a cover-up. For errors, always draw a line through the mistake and initial it.

Recordkeeping or documentation is written proof that your HACCP food safety system is in action and that you prepare, serve, and sell safe food. If a foodborne illness or outbreak were to occur, HACCP food safety records provide a *reasonable care defense* for your foodservice operation. The retail foodservice industry has a continuing legal duty to exercise the standards of reasonable care to anticipate, warn, and eliminate unreasonable risks of harm.

Records document the monitoring of the implementation of preventive controls and contain the actual values and observations obtained during monitoring. This information entered on records is based on the time the observations were made. These records are subject to verification using third-party organizations and their auditing systems.

Records that document the verification activities are used by the facility and by auditors conducting a food defense survey. These records shall include (1) the vulnerability assessment baseline survey, (2) a written food defense plan, (3) results of ongoing verification, and (4) documentation from the reassessments.

In contrast, all parties conducting business in the agricultural and/or food sector should require a written Food Defense SHARE plan. The Food Defense SHARE plan should be signed and dated by the most responsible individual on-site at your establishment. This signature ensures that the Food Defense SHARE plan has been accepted by the establishment and indicates its commitment to serving food that is both safe and secure. Records that document the training of each employee shall be retained according the Food Defense SHARE plan.

The recordkeeping for HACCP food safety is used in food defense during an investigation to provide a timeline as well as to determine the responsible party or potential witness to any alleged incident. Realize that some documents related to food defense must be password protected or secured under lock and key.

National Center for Food Protection and Defense

The leading organization for intentional contamination is the National Center for Food Protection and Defense (NCFPD). NCFPD was charged with leading the food and agriculture sector in developing a multidisciplinary and action-oriented research consortium focused on identifying vulnerabilities in the nation's food system. It is essential to take everything that you have learned in this chapter and apply it to the flow of food approach. This approach provides you with insight into how foods flow (i.e., product stream or processing steps): purchase, receive, store, prepare, cook, hold, cool, reheat, and serve, at any given time. Remember, the people are the key ingredient; that's why you need to have a strong strategy in your organization.

Now, let's check your understanding of this chapter by completing the following two activities: Study Questions and Apply Your Knowledge/Close Hold.

Hazards Study Questions

1. What are the seven attributes for CARVER+Shock?

   a. ______

   b. ______

   c. ______

   d. ______

   e. ______

   f. ______

   g. ______

2. What are the seven steps for conducting a CARVER+Shock analysis?

   a. ______

   b. ______

   c. ______

   d. ______

e. ______________________________

f. ______________________________

g. ______________________________

3. What are the seven HACCP principles?

a. ______________________________

b. ______________________________

c. ______________________________

d. ______________________________

e. ______________________________

f. ______________________________

g. ______________________________

4. What are the five food defense corrective action steps?

a. ______________________________

b. ______________________________

c. ____________________

____

____

d. ____________________

____

____

e. ____________________

____

____

5. What are the four factors that must be considered when conducting a vulnerability assessment throughout the flow of food?

a. ____________________

____

____

____

b. ____________________

____

____

____

c. ____________________

____

____

____

d. ____________________

____

____

____

APPLY YOUR KNOWLEDGE
(Close Hold Assignment)

1. What concepts could you suggest to your organization to prevent, eliminate, and reduce hazards?

2. How do you plan on implementing, monitoring, or testing your written food defense plan?

Congratulations! You have successfully completed *Hazards*, the second step in SHARE! Let's begin with Chapter 3, Accessibility . . .

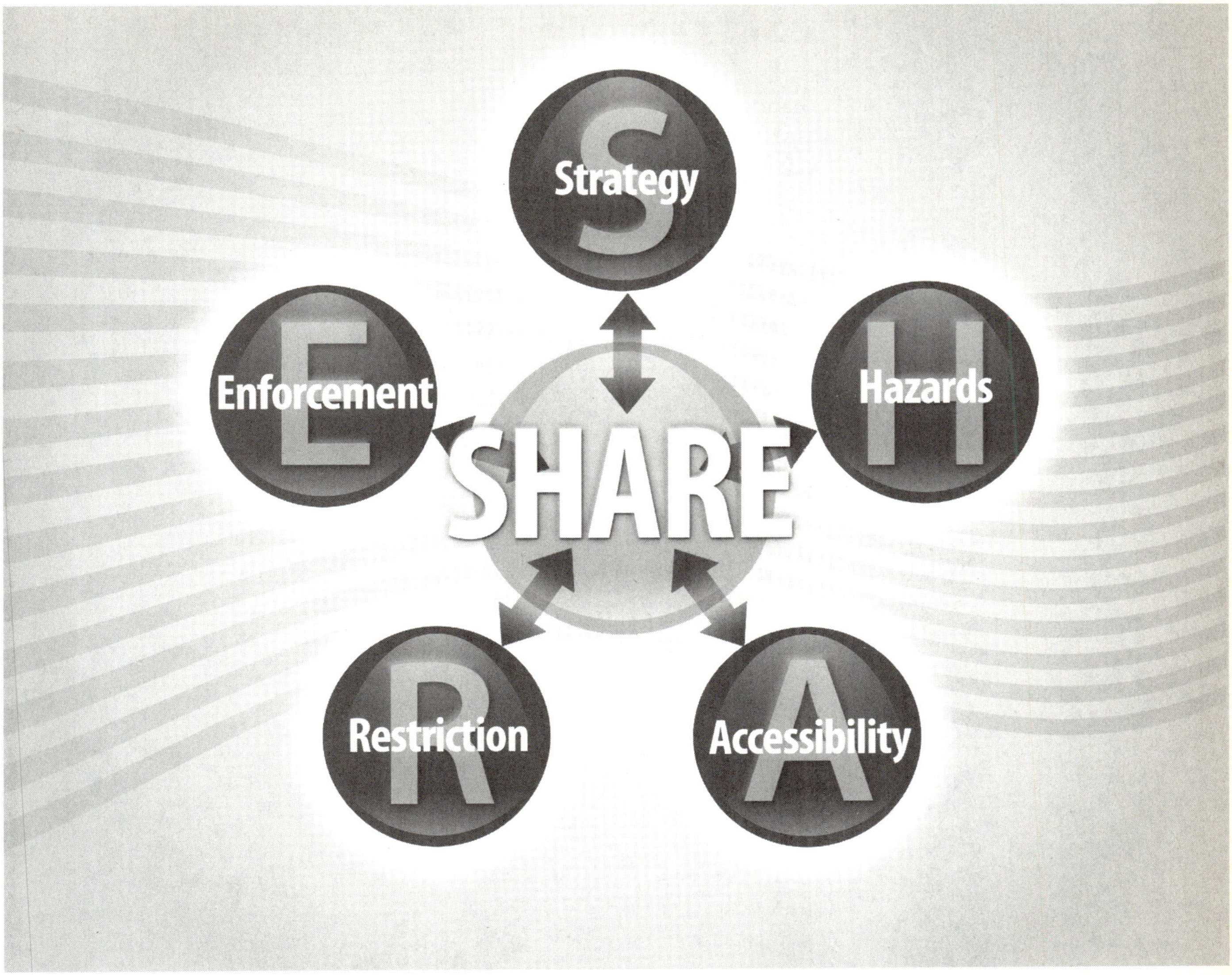

Chapter 3

Accessibility

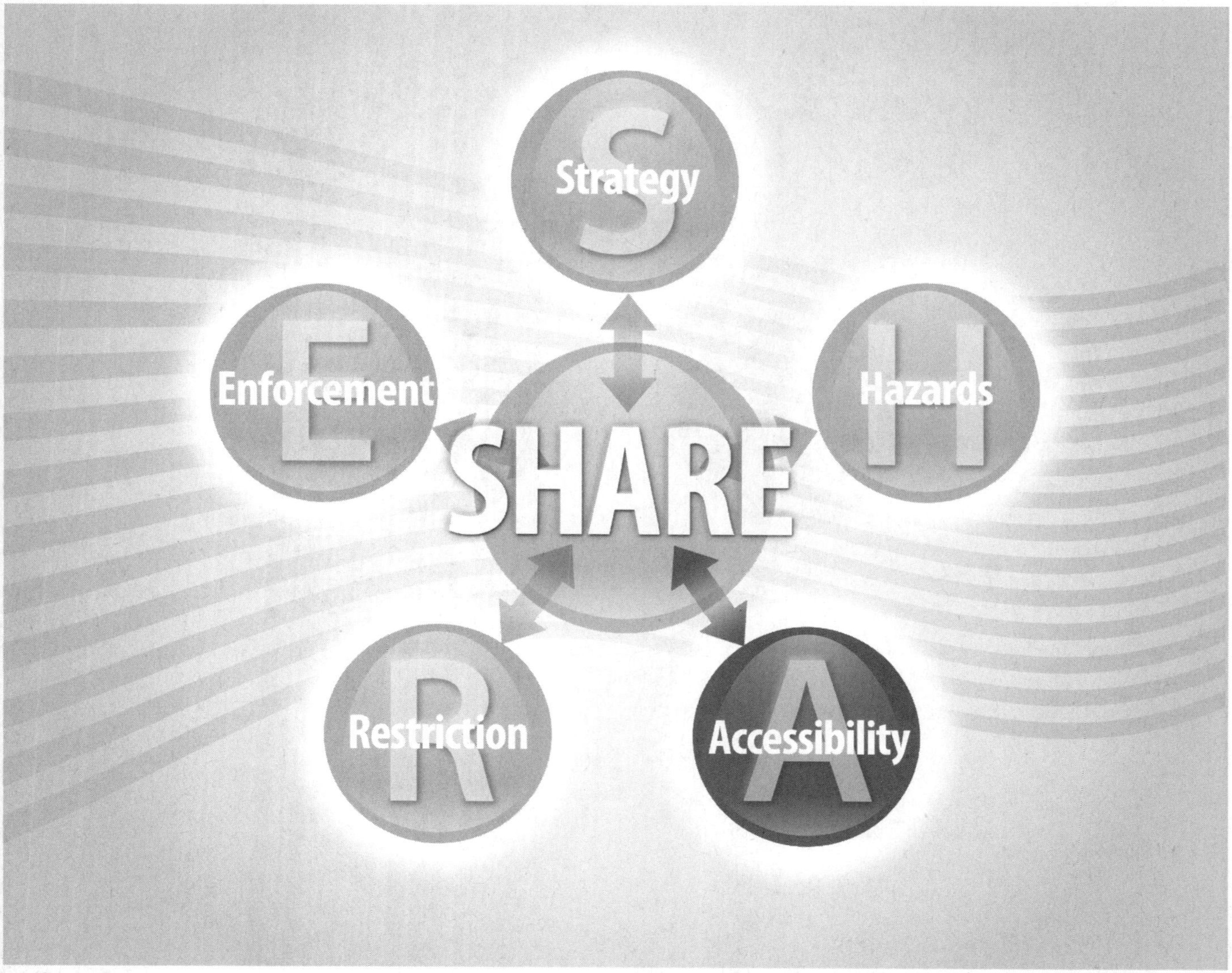

OVERVIEW

As mentioned in earlier chapters, protecting the food supply is not an option, but it *will* require you to develop a nonnegotiable food defense plan using the Food Defense SHARE principles. This chapter on accessibility is your opportunity to engage your workforce, through assignments, responsibilities, timelines, and milestones. Accessibility rights need to be determined for each aspect of the flow of food: purchasing, receiving, storing, preparing, cooking, holding, cooling, reheating, and serving. Any of these steps in the flow of food are potential entry points for hazards to intentionally be introduced into the product or processing stream. The prevention of intentional contamination of food as a nonnegotiable component of any plan of action needs to be an integral and significant part of your organization's culture as it pertains to protecting the food supply. Once your plan of action is developed, you will need to establish a food defense benchmark that is comparable to the best practices in your foodservice on a continual basis. Your plan of action should establish the critical limits, as required in a functional physical security protocol for benchmarking. It should also provide your employees with ongoing performance analysis of your Food Defense SHARE procedures, practices, and processes.

Accessibility rights need to be determined for each aspect of the flow of food: purchasing, receiving, storing, preparing, cooking, holding, cooling, reheating, and serving.

Food Defense **Accessibility** specifies access rights and management's control on information in specific areas to subsequently protect customers through vigilance, awareness, and hardening your physical security capabilities. Ultimately, if you develop the right strategy, identify the potential hazards, and use these accessibility techniques, you will be more able to deter or prevent an attack. You will be able to protect human life and ensure that your brand remains stable as well.

We are building upon SHARE. In the first chapter we first focused on strategy and supportive human resource systems. After developing the concepts of strategy and having a strong "people foundation," we gained an understanding of weapons used by foreign and domestic terrorists. These weapons are biological and chemical hazards that could be used to cause harm. The hazards chapter provided insight on using the CARVER+Shock methodology, vulnerability assessment, and how HACCP food safety helps you to develop procedures, practices, and processes to prevent intentional acts of terrorism. Now, let's begin with the learning objectives of the third step in Food Defense SHARE for accessibility.

LEARNING OBJECTIVES

In this chapter, you will learn to:

- Understand accessibility as it relates to Food Defense SHARE.
- Identify the difference between control and influence and how you have the ability to control and influence accessibility.
- Describe four basic types of motive.
- Recognize how motive is linked to accessibility.
- Explain five levels of accessibility.
- Compare motive and accessibility to global food defense events.

WHAT DO YOU CONTROL? WHAT DO YOU INFLUENCE?

Let's analyze your operation as it relates to Food Defense SHARE and accessibility. The term "control" means your ability to directly change the behaviors of others within your operation. The term "influence" refers to your power to leverage or sway the behaviors of others. Place a check in the appropriate box if you have direct **control** or

influence over leadership, vendors, auditing, food supply, human resources, operations, customers, and physical security.

FIGURE 3.1 Is this control or influence.

What Is Your Span of Control or Influence?	Control	Influence
Leadership		
1. Making food defense a priority.	☐	☐
2. Having food defense an important component to your business.	☐	☐
3. Redefining food defense in your company.	☐	☐
4. Incorporating food defense in your company.	☐	☐
Vendors		
5. Requiring vendors to have a food defense plan.	☐	☐
6. Verifying and validating vendor food defense plans.	☐	☐
7. Ensuring the integrity of food products received from vendors.	☐	☐
Auditing		
8. Conducting internal audits.	☐	☐
9. Incorporating food defense auditing requirements into your culture.	☐	☐
10. Hiring a third-party auditor to validate your food defense plan.	☐	☐
Food Supply		
11. Changing questionable food suppliers.	☐	☐
12. Traceability procedures.	☐	☐
13. Requiring food defense language in all contracts.	☐	☐

What Is Your Span of Control or Influence?	Control	Influence
Human Resources		
14. Updating your human resource process.	☐	☐
15. Hiring process to include background checks.	☐	☐
16. Hiring process to include food defense orientation.	☐	☐
17. Documenting proof of food defense training.	☐	☐
Operations		
18. Developing a food defense culture.	☐	☐
19. Creating a culture of interdependency.	☐	☐
20. Communicating food defense to your employees.	☐	☐
21. Documenting food defense.	☐	☐
22. Encourage reporting of suspicious activity.	☐	☐
23. Applying strategies throughout the flow of food.	☐	☐
24. Developing processes, practices, and procedures to protect your operation.	☐	☐
Customers		
25. Investigating customer complaints.	☐	☐
26. Developing a food defense hotline.	☐	☐
27. Developing a public announcement for your customers to report suspicious activity.	☐	☐
Physical Security		
28. Examining how food defense affects your physical security requirements.	☐	☐
29. Exploring how food defense overlaps with physical security and loss prevention.	☐	☐
30. Investigating how physical security and food defense relate to local law enforcement involvement.	☐	☐

- If you answered *control* to a statement, then the next step is to assign the name of a responsible person who is accountable for this job function. If you cannot assign a name, then how can you control it? Place the number here that you can both control and assign responsibility for that job function ______________________.
- If you answered *influence*, assign the name of the next best person who owns that job function. If you wanted to split or share the responsibility, who would you contact? Place the number here that you could not assign a name to the job function ______________________.

Reality Check

A Global Timeline of Food Defense Events

FIGURE 3.2 Food defense is a global issue.

1946 In Nuremberg, Germany, a group named NAKAM poisoned bread that was headed to an internment camp. Approximately 1,000 prisoners died as a result.

1984 In Dalles, Oregon, a group named the Rajneeshee cult contaminated salads and restaurants and a supermarket with a pathogen that caused 751 cases of *Salmonellosis.*

1989 In Philadelphia, Pennsylvania, grapes that were being exported from Santiago, Chile, to the United States and Japan were contaminated with cyanide. Only two grapes were found to be tainted with cyanide, but it was determined that cyanide deteriorates rapidly in grapes because of their high acidity. The fruit company that exported the grapes reported losses of $300 million, and U.S. importers lost over $100 million.

1994 In Dushanbe, Tajikstan, wine sold at a Russian military installation was delibe-rately contaminated with cyanide. Twenty-five persons became ill, and ten victims died.

1995 In Guangdong, China, two individuals were arrested for poisoning members and animals of their village with rat poison. As a result, approximately 163 people were sickened, 18 people died, and 4,000 animals perished.

2000 In Wakayama, Japan, a large pot of curry that was being prepared for a festival was sabotaged with arsenic and cyanide. Sixty-seven festivalgoers fell ill and four died.

2002 In Tangshan, Jiangsu Province, China, an individual used rat poison to taint a competitor's food products, resulting in the deaths of over forty-two persons.

2003 In Michigan, a disgruntled employee at a supermarket intentionally poisoned approximately 200 pounds of ground beef with a nicotine-based insecticide, resulting in about one hundred people becoming ill.

2003 In Yueyang, China, more than 200 people became ill after eating a breakfast tainted with rodenticide.

2003–2005 A series of incidents in various regions of Italy revealed that perpetrators had injected bleach, acetone, or ammonia into commercial drink containers. The attacks resulted in thirty-three deaths and significant financial losses for the drink industry.

2006 Three hundred police personnel became ill after they were poisoned at a cafeteria outside of Baghdad. Eleven deaths were reported, likely due to cyanide.

FIGURE 3.3 When there is a food defense event, the world is impacted.

Based on this timeline, the focus is on the need to understand that food safety and food defense are global issues. This timeline causes us to ask these questions: Could these events have been prevented? What were the motives behind the attacks? What mitigation strategies could have been used? What are the lessons learned from each one of these events? The key is to be vigilant and accountable in every aspect of your operation. The lessons learned from this timeline are useful as we develop a Food Defense SHARE plan. Once you learn how to incorporate accessibility strategies in your operation, you will understand how to strengthen any vulnerability that is exposed during the Food Defense SHARE course. It is critical to recognize that these food defense events have varying motives and different levels of accessibility. These will be explored later in this chapter. Let's look at some key terms to help understand accessibility and your role as a conscientious manager.

The key is to be vigilant and accountable in every aspect of your operation.

Key Terms Match Game

In the space provided, fill in the blank with the appropriate word that best fits the sentence as it relates to accessibility.

Stealth	**Processes**
Control	**Insider Attacker**
Influence	**Vigilance**
Procedures	**Accessibility**
Practices	**Need to Know**

(Continued)

1. ______________________________ is the act of paying close attention to and being watchful of your surroundings.

2. ______________________________ specifies access rights and management's control to information in specific areas to subsequently protect customers through vigilance, awareness, and hardening your physical security capabilities.

3. ______________________________ means your ability to direct or change the behaviors of others within your operation.

4. ______________________________ are guidelines, and customary or habitual behaviors for you, to perform in the most efficient manner.

5. ______________________________ is a trusted individual that gains your approval and access with ease to cause physical or mental harm.

6. ______________________________ refers to your power to leverage or sway the behaviors of others as it relates to your operation.

7. ______________________________ is acting in a covert or sneaky manner.

8. ______________________________ is a systematic series of coordinated tasks and activities to achieve a desired outcome in your operation.

9. ______________________________ is to selectively share information.

10. ______________________________ a particular course of action that you want your employees to achieve.

LEADERSHIP, SAFETY, AND SECURITY PROTOCOLS

What's Next? Where to Get Started with Accessibility?

Motive is the greatest difference between food safety and food defense events.

Food defense accessibility requires specifying access rights and controls to specific areas based on motive. A *motive* causes a person to act, whereas accessibility allows managers the access rights to resources ultimately for the prevention of disaster. Motive is the greatest difference between food safety and food defense events. In food safety incidents, there is no motive; and it is usually caused by an accidental or unintentional contamination. In general, a food safety event is not malice, nor is it intended to make people sick or cause death.

There is a history of foodborne illness outbreaks to use as a guide in understanding the potential impact as it relates to food defense. The impact would be increased exponentially if the motive were to intentionally contaminate the food by using a toxic agent. The motive is essential in understanding food defense. According to the World Health Organization, the following are the four categories for motive:

- disease, mass casualties, and death;
- economic and trade impact;

- impact on public health services; and
- social and political implications.

These motive categories have worldwide consequences. A human tragedy will create public outrage, especially when children are injured or killed. The negative impact of economic adulteration or disruption in trade is felt throughout the world. There are also major concerns for public health services being exhausted or overwhelmed due to disasters or emergencies. The social and political implications threaten civic order, create panic, and cause potential political destabilization.

There are eleven food defense-related events from our Global Timeline; each has one or more potential motives. As we look at accessibility levels, the timeline may help us understand what makes a terrorist target or attack the food supply to harm people or potentially destroy your company. The motives associated with these food defense events vary; some are obvious and some are not.

Accessibility must provide management and employees with the ability to secure the product and process for each step in the flow of food using a systematic approach. This systematic approach includes preventing, eliminating, and reducing hazards and applying procedures, practices, and processes.

Accessibility must provide management and employees with the ability to secure the product and process for each step in the flow of food using a systematic approach.

- Procedures are a particular *course of action* to achieve a desired result. Always start with the end in mind when you are creating your procedures.
- Practices are *guidelines, and customary or habitual behaviors* to perform in the most efficient manner.
- Processes are a systematic series of *coordinated tasks and activities* to achieve a desired outcome.

In your foodservice operation, accessibility needs to be managed and controlled while purchasing, receiving, storing, preparing, cooking, holding, cooling, reheating, and serving. Any of these steps in the flow of food are potential entry points for biological and chemical hazards to intentionally be introduced into the product or processing stream. Accessibility procedures help to prevent, eliminate, and reduce the chances of an attack occurring. Therefore, as you adopt the principles from the Food Defense SHARE certification program, you are contributing to the protection of our global food supply by making it less vulnerable.

The 2009 FDA Food Code referenced in this section demonstrates that the expectation already exists as part of what you do to run a safe operation. In the next exercise, you will see how establishing accessibility levels apply to the FDA Food Code. The three categories in the 2009 FDA Food Code are as follows: (1) a priority (P) item is a proven measure taken to directly control hazards associated with foodborne illness or injury, (2) a priority foundation (Pf) item requires specific actions to support one or more priority items and enable control of risk factors that contribute to foodborne disease or injury, and (3) a core (C) item is related to sanitation, facilities, and general maintenance.

FOOD DEFENSE SHARE ACCESSIBILITY LEVELS

Open—Easily Accessible

The most vulnerable and accessible area in your operation is defined as **open.** As per CARVER+Shock, this level would be *easily accessible*, such as if a target is outside the building, there is no perimeter fence, there are limited physical or human barriers or observation, and the attacker has relatively *unlimited access* to the target. At this point, an attack can be carried out without undue concern of detection. In addition, multiple sources of information concerning the facility and the target are easily available. In terms of control and influence, the operator needs a constant state of vigilance over accessibility in these critical areas. Following are two examples of easily accessible or *open* Food Defense SHARE level in a retail foodservice operation:

- In the front of the house, the self-service food items are open for your guests or customers to potentially contaminate. According to the 2009 FDA Food Code, "CONSUMER self-service operations such as buffets and salad bars shall be monitored by FOOD EMPLOYEES trained in safe operating procedures. [Pf]"
- Similarly, in the back of the house, an uncovered pot of food located on the stove may allow for intentional contamination. The buddy system increases the visibility of that area and would minimize the vulnerability.

High—Accessible

As per CARVER+Shock, the next most vulnerable area in your operation would be *accessible* to direct human contact to contaminate a product. This category is rated high in CARVER analysis if the target is inside the building, but in an unsecured part of your operation, where human observation and physical barriers are limited. An attacker has access to the target for *an hour or less.* This attack can be carried out, but requires the use of stealth. **Stealth** is behavior that includes acting in a covert or sneaky manner. Specific information about your operation may be easily available and known by the attacker. Following are two examples of highly accessible or *high* Food Defense SHARE level in a retail foodservice operation:

- In the front of the house, the operator has open coffeepots, and there is the potential for anyone to contaminate the coffee. Having air pots or a closed coffee system for the self-service coffee area would minimize this vulnerability. By changing the equipment to a closed coffee system, the risk would decrease, resulting in a low authorization level.
- In the back of the house, location of public restrooms in areas where there are limited employees and there is easy access to food products increases the risk to your operation, giving more opportunity for intentional contamination. Having public restrooms located in the front of the house for your guests or customers would lower the potential for accessibility to contaminate products. According to the 2009 FDA Food Code, "Toilet rooms shall be conveniently located and accessible to EMPLOYEES during all hours of operation. [C]"

Medium—Partially Accessible

As per CARVER+Shock, this level is *partially* accessible. Partially accessible would include having the attacker inside the building, but in a relatively unsecured, but busy, part of the operation. This area is under possible human observation; some physical barriers may be present and the contaminant may be disguised. Time *limitations are significant, 15–30 minutes.* Only general, nonspecific information is available on the facility and the target. Following are two examples of partially accessible or *medium* Food Defense SHARE level in a retail foodservice operation:

- In the front of the house, the bartender could intentionally contaminate the sanitizer sink behind the bar. This could harm your patrons and your brand if customers get sick. The use of cameras in the operation would serve as a *deterrent* if employees know they are under constant surveillance. According to the 2009 FDA Food Code, "EMPLOYEES are properly SANITIZING cleaned multiuse EQUIPMENT and UTENSILS before they are reused, through routine monitoring of solution temperature and exposure time for hot water SANITIZING, and chemical concentration, pH, temperature, and exposure time for chemical SANITIZING. [Pf]"

- In the back of the house, an employee could have sufficient amount of time, 15–30 minutes, to contaminate cases of product in storage. Again, the use of cameras or management self-inspections in the operation would serve as a *deterrent* if employees know they are under constant surveillance.

Low—Hardly Accessible

As per CARVER+Shock, this level is *hardly* accessible to an attack. A hardly accessible target or node is inside the building, but in a secured part of the operation. Human observation and physical barriers are used to detect suspicious activity. Accessibility to those areas is generally restricted to operators or authorized persons, and the contaminant must be disguised with *time limitations between 5–10 minutes.* There is limited general information available on the facility and the target. Following are two examples of hardly accessible or *low* Food Defense SHARE level in a retail foodservice operation:

- In the front of the house, while the server transports the food from the kitchen to the dining room, there is a potential to intentionally contaminate the food with an agent. Having a robust food defense culture such as trained conscientious managers, background checks, and open communication would improve detection.
- In the back of the house, there is a potential to accept contaminated product if the receiving person accepts an unlabeled and unmarked box from the delivery driver. In addition, the delivery driver assures the receiver that the content is safe but that the label fell off. The delivery driver then suggests that the product should be used immediately to avoid any confusion in storage. The receiver is influenced by the driver and then makes a poor decision to accept the product. Then receiver places the box directly into the preparation area for immediate use. This is an unacceptable practice and an example of how black market or potentially life-threatening food items enter into our food supply. In addition, this unsafe practice is not proper first in, first out (FIFO) food product rotation. It is essential to train your employees to reject unacceptable products and notify the person in charge immediately. According to the 2009 FDA Food Code, "EMPLOYEES are visibly observing FOODS as they are received to determine that they are from APPROVED sources, delivered at the required temperatures, protected from contamination, unADULTERED, and accurately presented, by routinely monitoring the EMPLOYEES' observations and periodically evaluating FOODS upon their receipt. [Pf]"

Closed—Not Accessible

As per CARVER+Shock, this level is *not* accessible to the introduction of a hazard. This closed authorization level focuses on physical barriers, alarms, and human observation to make it nonaccessible to direct human contact of the product. There is a defined means of intervention in place throughout the flow of food. An attacker can access the target, which are vulnerable points or steps, for *less than 5 minutes.* The attacker has no public information or training available concerning the target in your operation. Following are two not accessible or *closed* Food Defense SHARE level examples of a retail foodservice operation:

- In the front of the house, the attacker could compromise a closed ice cream or milkshake system during the time of replenishing the mix with new product. A comprehensive food defense human resource system including orientation and training for all employees to report suspicious activity is critical to prevent contamination. If the food is reworked, then the authorization level would not be closed. As previously discussed, rework runs the increased risk of potential contamination. Adding rework to this closed system requires that the authorization level increase because of this high-risk procedure of having direct human contact. In order to maintain a low authorization level, rework should not be a standard operating procedure.

- In the back of the house, the ice machine is a closed system until the door is opened and the ice is scooped. Ice is a ready-to-eat food that is often forgotten as an ingredient that can be contaminated. According to the 2009 FDA Food Code, "EMPLOYEES are preventing cross-contamination of READY-TO-EAT FOOD with bare hands by properly using suitable UTENSILS such as deli tissue, spatulas, tongs, single-use gloves, or dispensing EQUIPMENT. Pf" When designing your operation, the ice machine should be installed in a highly visible location where employee behaviors can be monitored closely.

Summary of Accessibility Levels

Food Defense SHARE Accessibility Levels	How difficult is it to physically access and egress from target?	How much time would a person have to access and egress a target point throughout the flow of food?
❑ Open	Easily Accessible (e.g., target is outside building and there is no perimeter fence). Limited physical or human barriers or observation. Attacker has relatively unlimited access to the target. Attack can be carried out using medium or large volumes of contaminant without undue concern of detection. Multiple sources of information concerning the facility and the target are easily available.	Unlimited time
❑ High	Accessible (e.g., target is inside building, but in unsecured part of facility). Human observation and physical barriers limited. Attacker has access to the target for an hour or less. Attack can be carried out with moderate to large volumes of contaminant, but requires the use of stealth. Only limited specific information is available on the facility and the target.	Less than 1 hour
❑ Medium	Partially Accessible (e.g., inside building, but in a relatively an unsecured, but busy, part of facility). Under constant possible human observation. Some physical barriers may be present. Contaminant must be disguised, and time limitations are significant. Only general, nonspecific information is available on the facility and the target.	15–30 minutes
❑ Low	Hardly Accessible (e.g., inside building in a secured part of the facility). Human observation and physical barriers with an established means of detection. Access generally restricted to operators or authorized persons. Contaminant must be disguised, and time limitations are extreme. Limited general information available on the facility and the target.	5–10 minutes
❑ Closed	Not Accessible. Physical barriers, alarms, and human observation are present. There are defined means of intervention in place. Attacker can access target for less than 5 minutes with all equipment carried in pockets. No useful publicly available information concerning the target.	Less than 5 minutes

Apply Motive and Accessibility to the Reality Check

Using the points already discussed and the accessibility summary chart, check the appropriate box that best applies to the motive and level of accessibility that was breached for each scenario. Keep in mind that you can have more than one motive but only one accessibility level.

Reality Check Scenario	Identify the Motive:	What If? Worst-Case?	Accessibility Level:
1. 1946 In Nuremberg, Germany, a group poisoned bread that was headed to an internment camp. Approximately 1,000 prisoners died as a result.	❑ Disease/death ❑ Economic/trade ❑ Public health ❑ Social/political	During preparation, the time limitations were less than 30 minutes, and the target was partially accessible.	❑ Open ❑ High ❑ Medium ❑ Low ❑ Closed
2. 1984 In Dalles, Oregon, a group named the Rajneeshee cult contaminated salads in restaurants and a supermarket with a pathogen, that caused 751 cases of *Salmonellosis.*	❑ Disease/death ❑ Economic/trade ❑ Public health ❑ Social/political	During serving, the operation was easily accessible; but there were limited physical or human barriers or observation.	❑ Open ❑ High ❑ Medium ❑ Low ❑ Closed
3. 1989 In Philadelphia, Pennsylvania, grapes that were being exported from Santiago, Chile, to the United States and Japan were contaminated with cyanide. Only two grapes were found to be tainted with cyanide, but it was determined that cyanide deteriorates rapidly in grapes because of their high acidity. The fruit company that exported the grapes reported losses of $300 million and U.S. importers lost over $100 million.	❑ Disease/death ❑ Economic/trade ❑ Public health ❑ Social/political	During purchasing, the attacker had unlimited access to the target.	❑ Open ❑ High ❑ Medium ❑ Low ❑ Closed
4. 1994 Wine sold at a Russian military installation was deliberately contaminated with cyanide. This incident resulted in 25 persons becoming ill and 10 victims died.	❑ Disease/death ❑ Economic/trade ❑ Public health ❑ Social/political	During receiving, the attack required less than 45 minutes.	❑ Open ❑ High ❑ Medium ❑ Low ❑ Closed
5. 1995 Two individuals were arrested for using rat poison. They poisoned members and animals of their village, which resulted in approximately 163 people becoming sick, 18 people dying, and 4,000 animals perishing.	❑ Disease/death ❑ Economic/trade ❑ Public health ❑ Social/political	During holding, the attacker had unlimited access to the target.	❑ Open ❑ High ❑ Medium ❑ Low ❑ Closed
6. 2000 In a large pot of curry that was being prepared for a festival was sabotaged with arsenic and cyanide. This attack resulted in 67 festivalgoers fell ill and four died.	❑ Disease/death ❑ Economic/trade ❑ Public health ❑ Social/political	During cooking, and reheating the operation was easily accessible; but there were limited physical or human barriers or observation.	❑ Open ❑ High ❑ Medium ❑ Low ❑ Closed
7. 2002 An individual used rat poison to taint a competitor's food products, resulting in the death of over 42 persons.	❑ Disease/death ❑ Economic/trade ❑ Public health ❑ Social/political	During cooling, there were some physical barriers, and less than 30 minutes.	❑ Open ❑ High ❑ Medium ❑ Low ❑ Closed

(Continued)

Reality Check Scenario	Identify the Motive:	What If? Worst-Case?	Accessibility Level:
8. 2003 In Michigan, a disgruntled employee at a supermarket intentionally poisoned approximately 200 pounds of ground beef with a nicotine-based insecticide, resulting in about 100 people becoming ill.	❑ Disease/death ❑ Economic/trade ❑ Public health ❑ Social/political	During storage, the target was easily accessible, outside the building or operation, and time was limited to 8 hours.	❑ Open ❑ High ❑ Medium ❑ Low ❑ Closed
9. 2003 In Yueyang, China, more than 200 people became ill after eating breakfast tainted with rodenticide.	❑ Disease/death ❑ Economic/trade ❑ Public health ❑ Social/political	During preparation, the attacker had access restricted to authorized persons-only, and the available time was less than 10 minutes.	❑ Open ❑ High ❑ Medium ❑ Low ❑ Closed
10. 2003–2005 A series of incidents in various regions of Italy revealed that perpetrators had injected bleach, acetone, or ammonia into commercial drink containers. The attacks resulted in 33 deaths and significant financial losses for the drink industry.	❑ Disease/death ❑ Economic/trade ❑ Public health ❑ Social/political	During serving, the attacker used stealth, and less than an hour.	❑ Open ❑ High ❑ Medium ❑ Low ❑ Closed
11. 2006 300 police personnel became ill after they were poisoned at a cafeteria outside of Baghdad. 11 deaths were reported, likely due to cyanide, by the Iraqi Environment Ministry.	❑ Disease/death ❑ Economic/trade ❑ Public health ❑ Social/political	During serving, the attacker had open access to the target, for an unlimited amount of time.	❑ Open ❑ High ❑ Medium ❑ Low ❑ Closed

Accessibility: A Shared Responsibility

Businesses need to learn how to incentivize and reward employees as they learn the "why?" behind the "what needs to be done?"

The Food Defense SHARE principle of accessibility must be a part of a multilayered approach to recognizing good practices, behaviors, and catching employees doing good things that are in line with your safety and security protocols. Businesses need to learn how to incentivize and reward employees as they learn the "why?" behind the "what needs to be done?" Ask yourself, how can I incorporate Food Defense SHARE principles in my operation today? The requirements for a Food Defense SHARE plan must address the flow of food and assessment of vulnerability, and ensure that your staff is trained in the belief that their sole purpose is to maintain customer trust and loyalty.

The goal of Food Defense SHARE is not to hinder access to critical areas, or to stop normal business practices from occurring."

The goal of Food Defense SHARE is not to hinder access to critical areas, or to stop normal business practices from occurring. Normal business practices need to focus on asset protection, building public trust, delivering a product where the consumer has confidence in the quality of the food. If loss prevention fails, and an intentional contamination occurs; then, the potential results are economic adulteration, mental or emotional injuries, and physical injuries or death. This is why accessibility is a shared responsibility from farm to fork.

Accessibility refers to developing a nonnegotiable plan of action to protect the food supply from an insider attack at any point within purchasing, receiving, storing, preparing, cooking, holding, cooling, reheating, and serving. Your accessibility plan of action should serve as a guide as you monitor the accessibility points for the products

FIGURE 3.4 Food Defense SHARE is exactly that—sharing the responsibility.

produced by your operation. The need to provide accessibility rights to your employees should be reflected in your processes, practices, and procedures as dictated in your Food Defense SHARE plan. A Food Defense SHARE plan should clearly define: "What does an adequate level of accessibility look like?" It needs to specify the role of management within each of the critical nodes. This will protect the safety and security of the food product throughout the flow of food. The training must also address the "Need to Know." The Need to Know should include guidance on information sharing, communication, specific vulnerabilities, and mitigation strategies.

As you think about what accessibility rights to authorize, you need to always keep in mind this question, what is the potential for any person to carry a sufficient amount of a biological or chemical agent within your operation? Also, where could someone intentionally introduce a hazard into the food? Accessibility looks closely at systems that support a proactive and positive culture of commitment for reducing, preventing, and eliminating hazards in our food.

Businesses need to promote a Food Defense SHARE culture that corrects behaviors and actions and responds to gaps in HACCP food safety. All members within your organization need to adopt a positive attitude about food safety and food defense. The best way to accomplish a Food Defense SHARE culture is by providing awareness training throughout your operation and to your most valuable asset—your people.

A robust food defense SHARE plan should allow for rewards or recognition to anyone that takes action once a suspicious activity is identified. This is an example of how to build a safe and secured culture that is self-sustaining. It also ensures that best practices are promoted and used throughout your operation.

One important thing to remember is that you should never focus on behaviors by way of stereotypes, especially profiling employees on the grounds of ethnicity or religious faith.

Accessibility is based on what goes into having a practical plan of action for better traceability, communication, human resource systems, and contractual language. Protecting your customers, employees, management, brand, and industry, and even global security is a core mission of a Food Defense SHARE plan. One important thing to remember is that you should never focus on behaviors by way of stereotypes, especially profiling employees on the grounds of ethnicity or religious faith. These tactics do not work in a Food Defense SHARE plan.

More importantly, you are responsible for learning how to assign your trusted employees to key workstations, which includes those pieces of equipment that are vulnerable. You should remember the CARVER+Shock methodology and Food Defense SHARE principles as you look at the worst-case scenarios at each workstation- as if the operator at that station is the inside attacker. Next, you should be able to recognize what vulnerabilities are present and which mitigation strategies are required to close physical security gaps that may exist. The need for widespread adoption of a Food Defense SHARE plan is paramount. The reality is that a terrorist threat or attack would negatively impact your livelihood and could destroy your brand.

In conclusion, as it relates to a Food Defense SHARE principle of accessibility, managers must also be well trained in controlling sensitive information, vulnerability assessment and mitigation strategies. Managers should be assigned specific access or accessibility rights to help facilitate the company's ability to produce their products.

Accessibility Study Questions

1. What are the five accessibility levels in Food Defense SHARE?

a. ______________________________

____

____

b. ______________________________

____

____

c. ______________________________

____

____

d. ______________________________

____

____

e. ______________________________

____

____

2. What are the four motives identified by the World Health Organization (WHO)?

a. ______________________________

____

____

____

b. ______________________________

____

____

____

c. ______________________________

____

____

____

d. ______________________________

____

____

____

3. What is the definition of (P), (Pf), and (C)?

(P) ______________________________

____

____

____

____

(Pf) ______________________________

____

____

____

____

(C) ______________________________

____

____

____

____

4. What is control? What do you control?

____

____

____

____

____

____

____

____

____

5. What is influence? What do you influence?

APPLY YOUR KNOWLEDGE
(Close Hold Assignment)

1. What concepts could you suggest to your organization to control accessibility?

2. How can you strengthen your food defense culture by focusing on accessibility?

Congratulations! You have successfully completed *Accessibility*, the third step in SHARE! Let's begin with Chapter 4, Restriction . . .

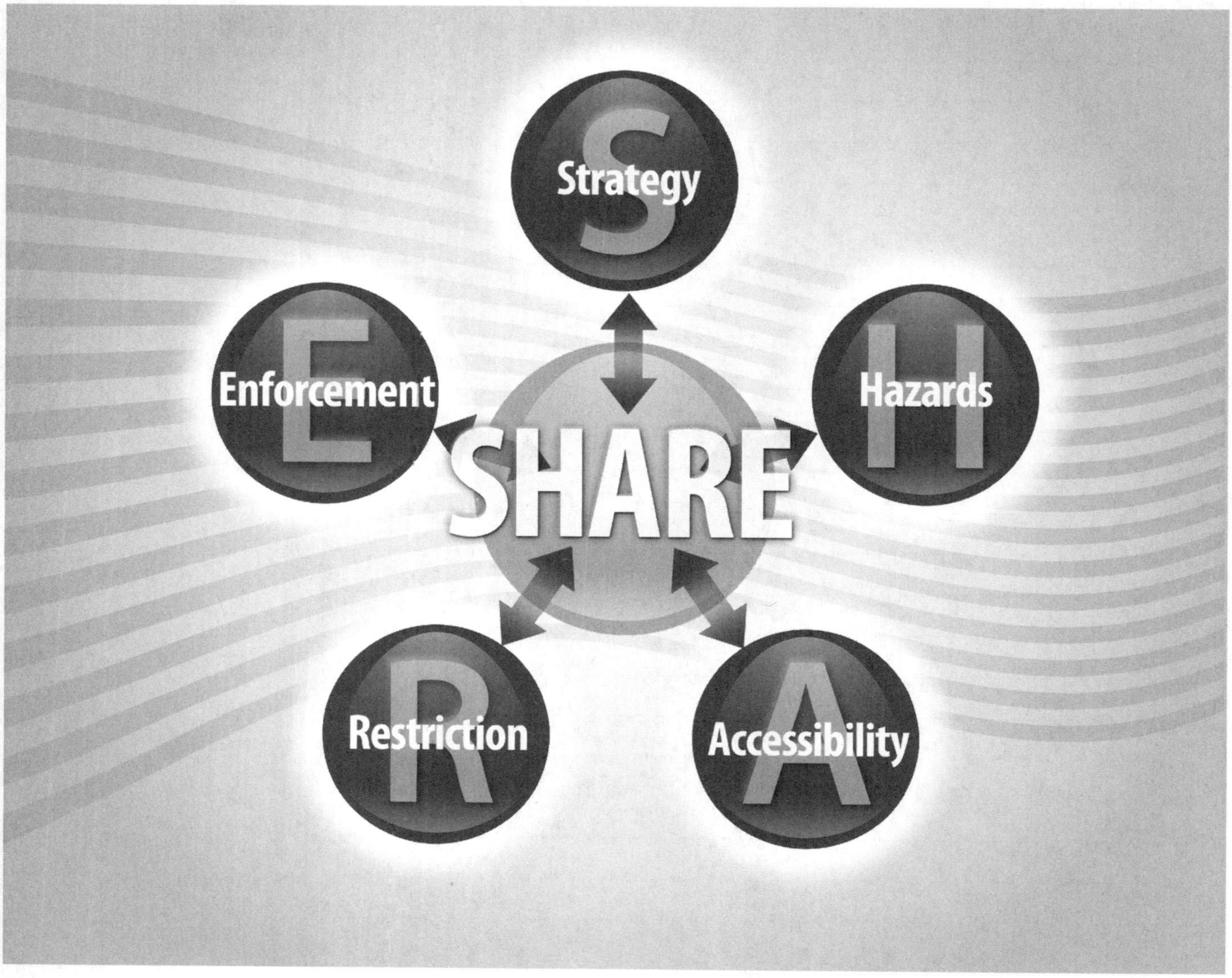

Chapter 4

Restriction

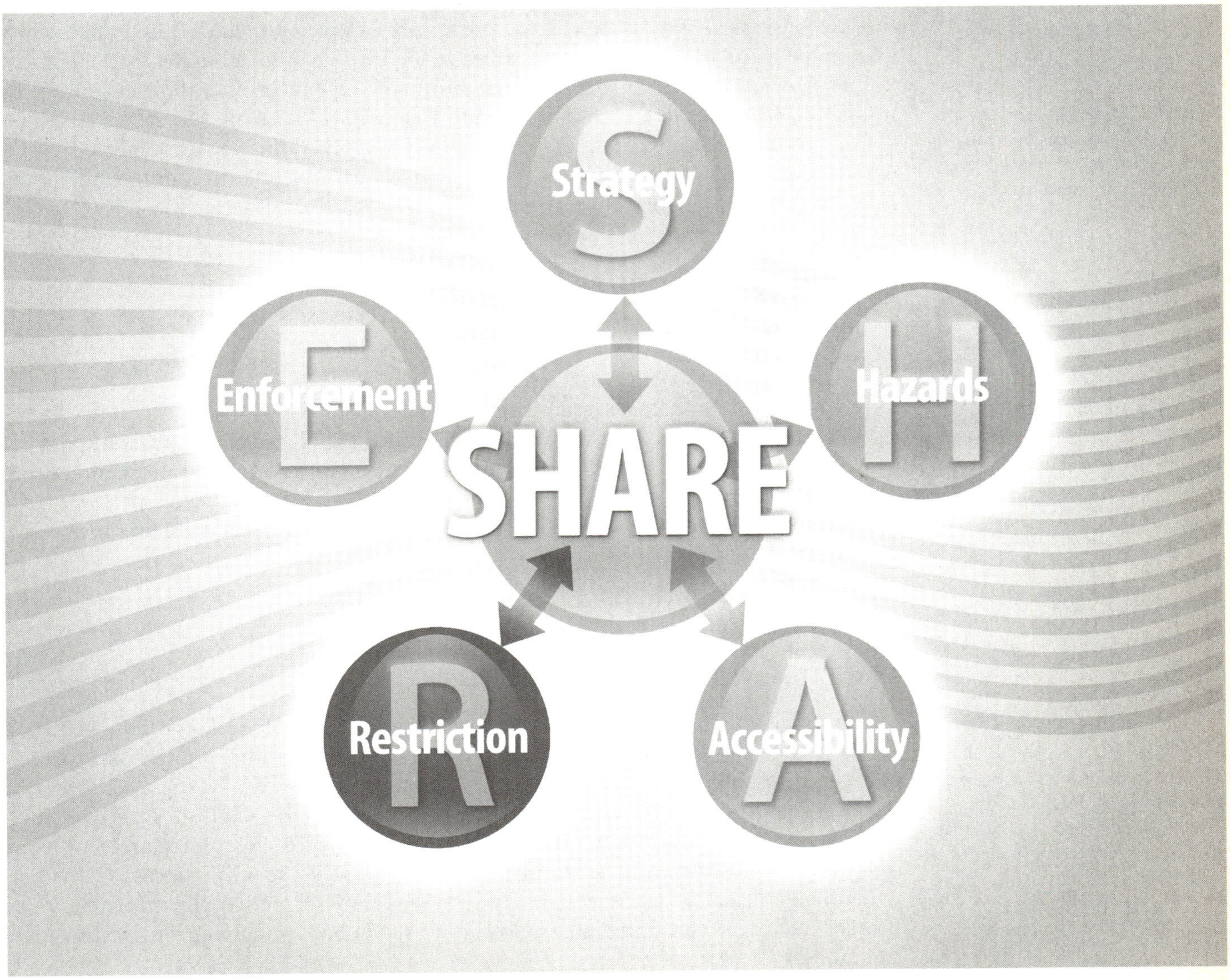

OVERVIEW

The need for continued *resiliency* within your organization begins with leadership taking responsibility for creating a Food Defense SHARE plan using principles of restriction. Restriction focuses on having a well-defined physical security protocol and related activities that include job duties for oversight, monitoring, and surveillance by key personnel within the boundaries of your operation. Restriction must be used as a proactive and preventive measure to defend against your vulnerabilities and guard your **brand shock value (BSV)**. Your brand's shock value is an expression used to measure the potential destruction of your brand or industry on a global scale. You need to know that an attack at your operation could potentially contribute to widespread fear and uncertainty.

Restriction must be used as a proactive and preventive measure to defend against your vulnerabilities and guard your **brand shock value** (BSV).

In this chapter, we will also apply vulnerability and restriction to the flow of food through various activities. We will use previously completed food defense vulnerability assessments and self-inspection checklists to identify, analyze, tabulate, and document vulnerabilities. Then, we will take this summary of your vulnerabilities and apply mitigation strategies to complete your Food Defense SHARE plan. Mitigation strategies are one of the most important steps that you can apply to the flow of food to prevent, eliminate, and reduce risk. Now, let's begin with the learning objectives of the fourth step in Food Defense SHARE—for restriction.

Mitigation strategies are one of the most important steps that you can apply to the flow of food to prevent, eliminate, and reduce risk.

LEARNING OBJECTIVES

In this chapter, you will learn to:

- Understand restriction as it relates to SHARE.
- Complete a food defense self-inspection checklist.
- Recognize brand shock value (BSV) and the psychological impact of CARVER+Shock as it relates to your operation, location, customers, and brand. This is the *Shock!*
- Apply restrictions as a proactive measure by scrutinizing vulnerabilities in the flow of food.
- Create a Food Defense SHARE plan.

EXERCISE: WHAT IS RESTRICTION?

Let's analyze your operation as it relates to Food Defense SHARE and how you have the ability to apply restriction. Restriction is based on imposing limitations and constraints throughout your operation. Restriction focuses on having a well-defined physical security protocol and related activities that include job duties for oversight, monitoring, and surveillance by key personnel within the boundaries of your operation. Keep in mind that food defense accessibility specifies access rights and control to critical areas. Accessibility provides management with information to protect customers through vigilance, awareness, and hardening your physical security capabilities.

Restriction focuses on having a well-defined physical security protocol and related activities that include job duties for oversight, monitoring, and surveillance by key personnel within the boundaries of your operation.

Use the self-inspection checklist below to conduct a food defense vulnerability assessment focusing on restriction needs of your operation to improve HACCP food safety and food defense. Answer *yes* or *no* to the following questions about your operation. Place a check in the appropriate box and complete the *Action Taken* section. Make sure that you do a thorough job because this exercise will be applied later in this chapter, as you learn to create a Food Defense SHARE plan. Let's complete the self-inspection checklist independently as this is a close hold assignment better known as "top secret."

FIGURE 4.1 Restriction safeguards are needed throughout your operations.

New Jersey Department of Health and Senior Services Food and Drug Safety Program

Food Defense Planning for Retail Food Establishments—A Self-inspection Checklist

Management	**Yes**	**No**	**Action Taken**
1. A food defense plan has been implemented at the facility.	☐	☐	
2. Food defense inspections of the facility are conducted.	☐	☐	
3. Appropriate management personnel have received training in food defense.	☐	☐	
4. The firm has an established food defense management team and/or food defense professional.	☐	☐	
5. Responsible personnel know whom to contact in the event of an emergency (both internally and externally).	☐	☐	
6. Security management program reviewed, verified, and revised, if necessary, at least annually.	☐	☐	
Personnel	**Yes**	**No**	**Action Taken**
7. Background checks as allowed by law are done on prospective employees.	☐	☐	
8. Employment applications are required.	☐	☐	
9. Employment references are checked.	☐	☐	
10. Personnel receive food defense training when they are hired.	☐	☐	
11. Only authorized personnel are allowed access to all parts of the facility.	☐	☐	
12. The facility has a system of positive identification and recognition for all personnel, for example, issuing photo ID.	☐	☐	
13. Customers are restricted to public areas.	☐	☐	
14. There are restrictions on personal items brought into the facility.	☐	☐	
15. Contractors are restricted to their work-required areas.	☐	☐	
16. Contractors and vendors are monitored while they are at the food facility.	☐	☐	
17. Employee sick leave policy encourages individuals to report illnesses and not work when they have gastrointestinal symptoms or a communicable disease.	☐	☐	
Property	**Yes**	**No**	**Action Taken**
18. A closed-circuit television system is used to monitor high-risk areas inside the facility.	☐	☐	
19. Security cameras are used to monitor the loading docks and exits.	☐	☐	
20. Hazardous chemical storage areas are secured.	☐	☐	
21. Doors opening onto the loading dock are kept locked when not in use.	☐	☐	
22. All truck shipments (incoming and outgoing) are monitored by foodservice employees.	☐	☐	

(Continued)

New Jersey Department of Health and Senior Services Food and Drug Safety Program (*Continued*)

Property	Yes	No	Action Taken
23. Products are inspected upon delivery.	☐	☐	
24. There is good lighting for all high-risk areas at the facility.	☐	☐	
25. High-risk areas are marked "employees only," and access is limited to employees who work in the area.	☐	☐	
26. There is a key control system for store keys.	☐	☐	
Products	**Yes**	**No**	**Action Taken**
27. Products are purchased from reputable established sources.	☐	☐	
28. Purchase records are maintained for product trace-back and recalls.	☐	☐	
29. Products arrive at the food facility in clean and secure transport vehicles.	☐	☐	
30. Products are never left unsupervised on the loading dock.	☐	☐	
31. Products are inspected for tampering prior to preparation or service.	☐	☐	
32. The facility has guidelines for handling product tampering incidents.	☐	☐	
33. Food items are prepared by personnel trained in food safety and food defense procedures.	☐	☐	
34. Potable water is used for rinsing and preparing food items.	☐	☐	
35. Salad bars and self-serve carts are closely monitored by staff to prevent contamination and product tampering.	☐	☐	

A self-inspection program can help you reduce the vulnerability of your food establishment to terrorist threats and criminal acts of product tampering.

A self-inspection program can help you reduce the vulnerability of your food establishment to terrorist threats and criminal acts of product tampering. In the past, food has been used as a vehicle to spread chemical and biological agents. Product tampering is a concern for the global food supply from farm to fork. Product tampering may also cause harm to customers and result in serious economic consequences for the foodservice establishment.

A self-inspection program is an important preventive measure that will improve food defense at your facility. This self-inspection checklist is a guide to help you evaluate your facility. This is easily customized for convenience stores, supermarkets, restaurants, casinos, military, schools, senior care facilities, airports, institutional foodservice, cruise ships, contract foodservice, hotels, commissaries, hospitals, distribution centers, colleges, and universities. Please modify this checklist to meet the circumstances of your operation. Remember that a Food Defense SHARE plan will help protect your customers, your employees, and your business.

Reality Check

New Hampshire: DHHS Issues Warning of Fraudulent Food Inspectors, June 17, 2010

Concord, NH—The New Hampshire Department of Health and Human Services (DHHS), Division of Public Health Services (DPHS), Food Protection Section is advising restaurant owners and operators to be wary of callers misrepresenting themselves as New Hampshire food inspectors.

"We have been alerted to a number of complaints from concerned restaurant owners from various regions of the State who have received calls from someone posing as a health inspector," said DHHS Public Health Director Dr. José Montero. "The fake inspector attempts to schedule an inspection at the restaurant and threatens to impose fines if they don't book an appointment. Our Food Protection Inspectors conduct unannounced inspections and would not be calling to schedule appointments for routine inspections. They also do not threaten enforcement actions or ask for confidential information over the phone."

FIGURE 4.2 Valued partnerships between local, state, tribal, federal, and international regulatory authorities, industry, and academia are engaged in food defense.

Officials conducting restaurant inspections will also have proper photo identification from the State of New Hampshire. Foodservice establishments are advised not to provide sensitive or confidential information over the telephone.

The scam is reportedly targeting restaurant owners here in New Hampshire and across the country.

Restaurant owners and operators are advised to immediately report suspicious or fraudulent calls to local law enforcement officials. If you have any questions or concerns, please contact the DHHS Food Protection Section at 603-271-4589. You can also call the New Hampshire Department of Justice Consumer Protection Hotline at 1-888-468-4454 between 9 am and 3 pm.

Educational materials for food establishment owners and managers to increase employee awareness of suspicious behavior are available at the following Web sites: http://www.fda.gov/food/fooddefense/training/alert/default.htm; http://www.fda.gov/Food/FoodDefense/Training/ucm135038.htm

---

Source: http://www.dhhs.state.nh.us/DHHS/PIO/LIBRARY/Press+Release/fraudulentinspectors.htm

In the above article, the focus is on restricting access and confidential information by imposing limitations and constraints throughout your operation in order to decrease the impact of your brand shock value. Foodservice retail operators should use Food Defense SHARE in preventing similar scenarios from occurring. If this situation is allowed, then an intentional contamination could occur and potentially lead to disaster. Throughout this entire book, we focus on awareness, accessibility, vulnerability, shock, and restriction. It is critical to restrict access to sensitive or confidential information *at all times.* Let's look at some key terms to understand restriction and how it applies to your operation.

Key Term Match Game

Answer the questions below by filling in the blank with the appropriate word that best fits the sentence as it relates to the restriction protocols of your organization.

Restriction	**Restrict**
Rules	**Breach**
Limitations	**Protocol**
Guidelines	**Vulnerability**
Constraints	**Susceptibility**

1. ______________________________
is the ease of accomplishing a desired outcome; and refers to areas of weakness in communication, infrastructure, physical building, equipment, and, more specifically, to the potential exposure to hazards.

2. ______________________________
is a gap, infraction, or violation of standard operating procedures.

3. ______________________________
is a course of action.

4. ______________________________
is a formal principle or regulation.

5. ______________________________
is a written description that outlines the steps necessary to ensure operational validity based on the best science and methodology.

6. ______________________________
is based on imposing limitations and constraints throughout your operation.

7. ______________________________
is keeping within certain limits.

8. ______________________________
is the process for controlling and protecting critical information or areas.

9. ______________________________
is restricting access and actions.

10. ______________________________
is a weakness or sensitivity allowing an action to be performed. This can be a characteristic that makes the global food supply vulnerable.

LEADERSHIP, SAFETY, AND SECURITY PROTOCOLS

What's Next? Where to Get Started?

Restriction follows the flow of food in order to analyze vulnerability and the brand shock value of your operation. As a result of this firsthand knowledge of your retail foodservice operation, restriction provides you with the keys to developing a food defense plan. Hiding an illegal practice, covering it up, or contributing to any terror, fear, or retaliation is a crime. Food defense awareness is a community activity where you have the freedom to execute your civic duty to assist in the moral and ethical order of society. The concept of "If You See Something, Say Something" should empower you to take action in order to protect human life, the global food supply and your brand. This is a preventive measure you can use to expose vulnerabilities in your operation. Vulnerability refers to areas of weakness in communication, infrastructure, physical building, equipment, and, more specifically, to the potential exposure to hazards.

Food defense awareness is a community activity where you have the freedom to execute your civic duty to assist in the moral and ethical order of society.

As per CARVER+Shock, vulnerability is a measure of the ease with which biological and/or chemical threat agents can be introduced in quantities sufficient to achieve the attacker's purpose. The brand shock value is an expression used to measure the potential destruction of your brand, industry, and global economics that could potentially contribute to widespread fear and uncertainty. The shock attributes in CARVER+Shock methodology are "a combined measure of the health, psychological, and collateral national economic impacts of a successful attack on the target system. In addition, Shock is considered on a national level. The psychological impact will be increased if there are a large number of deaths or the target has historical, cultural, religious or other symbolic significance. Mass casualties are not required to achieve widespread economic loss or psychological damage. Collateral economic damage includes such items as decreased national economic activity, and increased unemployment in collateral industries. Psychological impact will be increased if victims are members of sensitive subpopulations such as children or the elderly."

The brand shock value is an expression used to measure the potential destruction of your brand, industry, and global economics that could potentially contribute to widespread fear and uncertainty.

In summary, vulnerability, accessibility, and brand shock value are determined by the following factors, but are not limited to them:

- short shelf life
- toxicity of an agent
- ability to detect suspicious activity
- ease of accessing key workstations
- uniformly mixing of an agent
- visibility of employees in critical areas
- conditions in the surrounding area/environment
- amount of training needed to identify the target node
- time constraints to accomplish an attack

Knowing the characteristics of vulnerability decreases the risk of an attack. After you identify the vulnerabilities in your operation, you can decide what will go into your food defense plan in terms of restriction. Let's review the characteristics that make food more vulnerable.

The food in your operation is at a *higher risk* if you use:

- ready-to-eat, preprepared food because you are trusting your supplier;
- complex recipes that move through the temperature danger zone two times and require multiple preparation steps because human contact is increased;
- elaborate foods composed of ingredients from many sources that may have more potential for contamination;
- specialty-type foods, which include iconic, symbolic, and name brand products that are globally distributed; and
- short shelf life foods due to these products being consumed more rapidly.

It also depends on factors such as:

- whether a particular commodity is subject to extensive testing or not;
- distance food travels;
- multiple uses of an ingredient in other foods;
- food consumption patterns for various foods; and
- global significance for the food products or venue.

FIGURE 4.3 Macaroni Salad. What are the vulnerabilities?

Using vulnerability along with the shock attribute provides you with the psychological impact of an intentional contamination of food. Resiliency is magnified by the need to prevent the psychological impact of an attack which is why restriction is critical. The terrorist measures success based on the economical and psychological impacts of increased fear or death.

Your next task is to review a series of scenarios related to enhancing your sense of awareness of the vulnerabilities in the flow of food. Identify vulnerabilities and apply mitigation strategies using the recipes provided by explaining why each workstation or node may be more susceptible. Therefore, when a well-defined mitigation is determined, it should remove any potential threat. The following three recipes are examples within the flow of food, which are purchase, receive, store, prepare, cook, hold, cool, reheat, and serve:

1. Three Bean Salad (purchase, receive, and store)
2. Lasagna with Ground Beef (prepare, cook, and hold)
3. Chili con Carne with Beans (cool, reheat, and serve)

Similar to HACCP, these scenarios are divided into three categories based on the number of times the food product moves through the temperature danger zone. These three categories are no-cook/simple, same-day, and complex.

According the best practices in HACCP Food Safety, the temperature danger zone is 41°F to 135°F (or 5°C to 57.2°C). A no-cook/simple recipe does not move through the temperature danger zone. There is no cooking, or kill step involved. A same-day recipe means the product is prepared for same-day service or has some same-day cooking nvolved, where the food will pass through the temperature danger zone one time before it is served or sold. A complex recipe moves food through the temperature danger zone two times, and it is prepared, cooled, and stored and can be reheated. The following activity uses all three categories:

- Activity 1: No-cook/simple recipe
- Activity 2: Same-day recipe
- Activity 3: Complex recipe

Remember to only discuss what happens at each step in the flow of food, not what may happen before or after each designated workstation or activity step. You need to think about the worst-case scenario for the vulnerability and apply restriction as a mitigation step for each example. Let's get started.

NO-COOK/SIMPLE RECIPE: PURCHASE, RECEIVE, AND STORE

Activity 1: Three Bean Salad

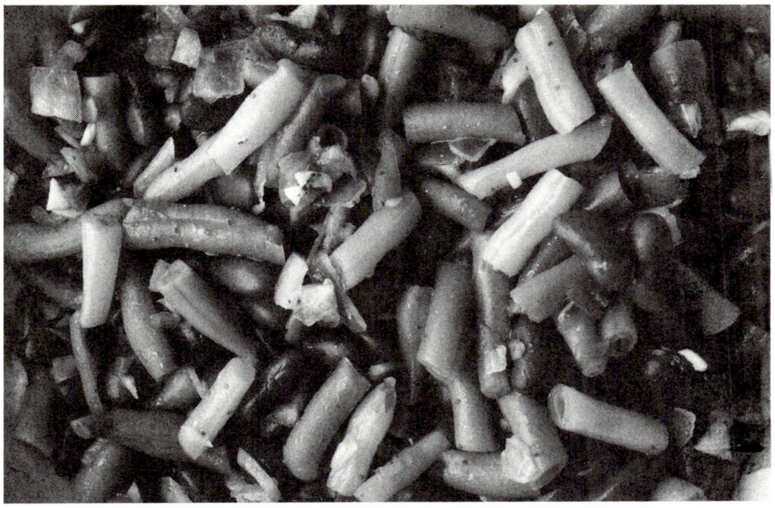

FIGURE 4.4 Three Bean Salad.

What are the vulnerabilities?

Three Bean Salad

Vegetable | Salads and Salad Dressings | E-11

Ingredients	50 Servings		100 Servings		Directions
	Weight	Measure	Weight	Measure	
Canned kidney beans, chilled, drained	1 lb 2 oz	1 ½ cups 1 Tbsp (¼ No. 10 can)	2 lb 4 oz	3 cups 2 Tbsp (½ No. 10 can)	1. Rinse kidney beans in cold water and drain well.
Canned wax beans, chilled, drained	15 oz	2 ¾ cups (¼ No. 10 can)	1 lb 14 oz	1 qt 1 ½ cups (½ No. 10 can)	2. Combine kidney beans, wax beans, green beans, onions, and green peppers (optional).
Canned cut green beans, chilled, drained	2 lb 5 ½ oz	1 qt 3 ½ cups (⅔ No. 10 can)	4 lb 11 oz	1 gal (1 ¼ No. 10 cans)	
*Fresh white or red onions, chopped	3 oz	½ cup	6 oz	1 cup	
*Fresh green peppers, chilled, chopped (optional)	6 ½ oz	1 ¼ cups	13 oz	2 ½ cups	
Vegetable oil		⅔ cup		1 ⅓ cups	3. Combine vegetable oil, vinegar, sugar, basil, pepper, and granulated garlic. Mix until well blended.
White vinegar		⅔ cup		1 ⅓ cups	
Sugar	2 ¼ oz	⅓ cup	4 ½ oz	⅔ cup	
Dried basil		1 Tbsp		2 Tbsp	
Ground black or white pepper		1 tsp		2 tsp	
Granulated garlic		1 tsp		2 tsp	
					4. Pour dressing over beans. Toss lightly to combine and coat evenly. Spread 5 lb 12 ½ oz (approximately 3 qt 1 cup) into each shallow pan (12" x 20" x 2 ½") to a product depth of 2" or less. For 50 servings, use 1 pan. For 100 servings, use 2 pans.
					5. CCP: Cool to 41° F or lower within 4 hours. Refrigerate until ready to serve.
					6. Mix lightly before serving. Portion with No. 16 scoop (¼ cup).

Comments:
*See Marketing Guide.

Marketing Guide for Selected Items

Food as Purchased for	50 Servings	100 Servings
Mature onions	4 oz	8 oz
Green peppers	9 oz	1 lb 2 oz

Three Bean Salad

Vegetable | Salads and Salad Dressings | E-11

SERVING:	YIELD:		VOLUME:	
¼ cup (No. 16 scoop) provides ¼ cup of vegetable.	**50 Servings:**	about 5 lb 12 ½ oz	**50 Servings:**	about 3 quarts 1 cup 1 pan
	100 Servings:	about 11 lb 9 oz	**100 Servings:**	about 1 gallon 2 ½ quarts 2 pans

Edited 2004

Nutrients Per Serving

Calories	47	**Saturated Fat**	0.42 g	**Iron**	0.45 mg
Protein	0.93 g	**Cholesterol**	0 mg	**Calcium**	13 mg
Carbohydrate	4.56 g	**Vitamin A**	92 IU	**Sodium**	110 mg
Total Fat	2.99 g	**Vitamin C**	1.6 mg	**Dietary Fiber**	1.0 g

What restrictions can be applied?

FIGURE 4.5 Dry Storage.

SAME-DAY RECIPE: PREPARE, COOK, AND HOLD

Activity 2: Lasagna with Ground Beef

FIGURE 4.6 Lasagna.

What are the vulnerabilities?

Lasagna with Ground Beef

Ingredients	50 Servings		100 Servings		Directions
	Weight	Measure	Weight	Measure	
Raw ground beef (no more than 20% fat)	3 lb 4 oz		6 lb 8 oz		1. Brown ground beef. Drain. Continue immediately. Add onions and granulated garlic. Cook for 5 minutes.
*Fresh onions, chopped OR Dehydrated onions	3 lb OR 9 oz	2 qt OR 1 qt ½ cup	6 lb OR 1 lb 2 oz	1 gal OR 2 qt 1 cup	
Granulated garlic		2 Tbsp		¼ cup	
Ground black or white pepper		1 tsp		2 tsp	2. Add pepper, parsley, canned tomatoes, tomato paste, water, and seasonings. Heat to boiling, uncovered. Remove from heat.
Dried parsley		¼ cup		½ cup	
Canned diced tomatoes, with juice	4 lb 4 oz	2 qt ¼ cup (⅔ No. 10 can)	8 lb 8 oz	1 gal ½ cup (1 ⅓ No. 10 cans)	
Canned tomato paste	1 lb 12 oz	3 cups 2 Tbsp (¼ No. 10 can)	3 lb 8 oz	1 qt 2 ¼ cups (½ No. 10 can)	
Water		3 qt		1 gal 2 qt	
†Seasonings Dried basil Dried oregano Dried marjoram Dried thyme		 3 Tbsp 3 Tbsp 1 Tbsp 1 tsp		 ¼ cup 2 Tbsp ¼ cup 2 Tbsp 2 Tbsp 2 tsp	
Enriched lasagna noodles, uncooked (at least 0.78 oz each)	2 lb 12 oz	56 each	5 lb 8 oz	112 each	3. Assemble ingredients as follows: In steamtable pans (12" x 20" x 2 ½") which have been lightly coated with pan release spray. For 50 servings, use 2 pans. For 100 servings, use 4 pans. For each pan 1st layer-1 qt ½ cup sauce 2nd layer-14 uncooked noodles lengthwise 3rd layer-1 qt ½ cup sauce 4th layer-6 ¼ oz (1 ½ cups 1 Tbsp) cheese blend and 9 ½ oz (2 ¼ cups 2 Tbsp) mozzarella cheese 5th layer-14 uncooked noodles crosswise 6th layer-1 qt ¾ cup sauce 7th layer-6 ¼ oz (1 ½ cups 1 Tbsp) process American cheese and 9 ½ oz (2 ¼ cups 2 Tbsp) mozzarella cheese

Lasagna with Ground Beef

Cheese blend of American and skim milk cheeses, shredded	1 lb 9 oz	1 qt 2 ¼ cups	3 lb 2 oz	3 qt ½ cup	
Lite mozzarella cheese, shredded	2 lb 6 oz	2 qt 1 ½ cups	4 lb 12 oz	1 gal 3 cups	
					4. Tightly cover pans.
					5. Bake: Conventional oven: 350° F for 1 ¼ -1 ½ hours Convection oven: 325° F for 45 minutes CCP: Heat to 165° F or higher for at least 15 seconds.
					6. Remove pans from oven. Uncover. Let stand for 15 minutes before serving.
					7. CCP: Hold for hot service at 135° F or higher. Cut each pan 5 x 5 (25 pieces per pan).

Comments:
*See Marketing Guide.

†Italian Seasoning Mix (see G-01, Sauces, Gravies, and Seasoning Mixes) may be used to replace these ingredients. For 50 servings, use ¼ cup 3 Tbsp 1 tsp Italian Seasoning Mix. For 100 servings, use ¾ cup 2 Tbsp 2 tsp Italian Seasoning Mix.

Marketing Guide for Selected Items

Food as Purchased for	50 Servings	100 Servings
Mature onions	3 lb 7 oz	6 lb 14 oz

SERVING:	YIELD:		VOLUME:	
1 piece provides 2 oz equivalent meat/meat alternate, ⅜ cup of vegetable, and ¾ serving of grains/breads.	**50 Servings:**	about 22 lb 12 oz	**50 Servings:**	2 steamtable pans
	100 Servings:	about 45 lb 8 oz	**100 Servings:**	4 steamtable pans

Tested 2004

Lasagna with Ground Beef

Meat/Meat Alternate-Vegetable-Grains/Breads | Main Dishes | D-25

Nutrients Per Serving					
Calories	269	**Saturated Fat**	4.69 g	**Iron**	2.65 mg
Protein	19.88 g	**Cholesterol**	34 mg	**Calcium**	304 mg
Carbohydrate	28.35 g	**Vitamin A**	910 IU	**Sodium**	406 mg
Total Fat	8.70 g	**Vitamin C**	16.3 mg	**Dietary Fiber**	2.3 g

What restrictions can be applied?

COMPLEX RECIPE: COOL, REHEAT, AND SERVE

Activity 3: Chili con Carne with Beans

FIGURE 4.7 Chili.

Chili con Carne with Beans

Ingredients	50 Servings		100 Servings		Directions
	Weight	Measure	Weight	Measure	
Raw ground beef (no more than 20% fat)	7 lb		14 lb		1. Brown ground beef. Drain. Continue immediately.
*Fresh onions, chopped OR Dehydrated onions	14 oz OR 2 ½ oz	2 ⅓ cups OR 1 ¼ cups	1 lb 12 oz OR 5 oz	1 qt ⅔ cup OR 2 ½ cups	2. Add onions, granulated garlic, green pepper (optional), pepper, chili powder, paprika, onion powder, and ground cumin. Cook for 5 minutes.
Granulated garlic		1 Tbsp 1 ½ tsp		3 Tbsp	
*Fresh green pepper, chopped (optional)	8 oz	1 ½ cups 2 Tbsp	1 lb	3 ¼ cups	
Ground black or white pepper		2 tsp		1 Tbsp 1 tsp	
Chili powder		3 Tbsp		¼ cup 2 Tbsp	
Paprika		1 Tbsp		2 Tbsp	
Onion powder		1 Tbsp		2 Tbsp	
Ground cumin	1 oz	¼ cup	2 oz	½ cup	
Canned diced tomatoes, with juice	3 lb 3 oz	1 qt 2 ¼ cups (½ No. 10 can)	6 lb 6 oz	3 qt ½ cup (1 No. 10 can)	3. Stir in tomatoes, water, and tomato paste; mix well. Bring to boil. Reduce heat. Cover. Simmer slowly, stirring occasionally until thickened, about 40 minutes.
Water		2 qt 1 cup		1 gal 2 cups	
Canned tomato paste	1 lb 12 oz	3 cups 2 Tbsp (¼ No. 10 can)	3 lb 8 oz	1 qt 2 ¼ cups (½ No. 10 can)	
Canned pinto or kidney beans, drained OR *Dry pinto or kidney beans, cooked (see Special Tip)	3 lb 6 oz OR 2 lb 4 oz	1 qt 3 ½ cups (½ No. 10 can) OR 1 qt 2 cups	6 lb 12 oz OR 4 lb 8 oz	3 qt 3 cups (1 No. 10 can) OR 3 qt	4. Stir in beans. Cover and simmer. Stir occasionally. CCP: Heat to 155° F or higher for 15 seconds. OR If using previously cooked and chilled beans: CCP: Heat to 165° F or higher for at least 15 seconds.
					5. Pour into serving pans.
					6. CCP: Hold for hot service at 135° F or higher. Portion with 4 oz ladle (½ cup).
Reduced fat Cheddar cheese, shredded (optional)	1 lb 8 oz	1 qt 2 cups	3 lb	3 qt	7. Garnish with cheese (optional).

Chili con Carne with Beans

Comments:
*See Marketing Guide.

Marketing Guide for Selected Items

Food as Purchased for	50 Servings	100 Servings
Mature onions	1 lb	2 lb
Green peppers	11 oz	1 lb 6 oz
Dry pinto beans, dry OR Dry kidney beans	1 lb OR 1 lb	2 lb OR 2 lb

SERVING:

½ cup (4 oz ladle) provides 2 oz equivalent meat/meat alternate and ⅜ cup of vegetable.

YIELD:

50 Servings: about 16 lb 4 oz

100 Servings: about 32 lb 8 oz

VOLUME:

50 Servings: about 1 gallons 2 ¼ quarts

100 Servings: about 3 gallons 2 cups

Tested 2004

Special Tip:
SOAKING BEANS
Overnight method: Add 1 ¾ qt cold water to every 1 lb of dry beans. Cover and refrigerate overnight. Discard the water. Proceed with recipe.

Quick-soak method: Boil 1 ¾ qt of water for each 1 lb of dry beans. Add beans and boil for 2 minutes. Remove from heat and allow to soak for 1 hour. Discard the water. Proceed with recipe.

COOKING BEANS
Once the beans have been soaked, add ½ tsp salt for every lb of dry beans. Boil gently with lid tilted until tender, about 2 hours.

Use hot beans immediately.
CCP: Hold for hot service at 135° F.
OR
Or, chill for later use.
CCP: Cool to 70° F within 2 hours and to 41° F or lower within an additional 4 hours.
Or, chill for later use.
CCP: Cool to 70° F within 2 hours and to 41° F or lower within an additional 4 hours; and CCP: Reheat to 165 ° F for 15 seconds within 2 hours. Serve

1 lb dry pinto beans = about 2 ⅜ cups dry or 5 ¼ cups cooked beans.

Variation:
A. Chili con Carne without Beans

50 servings: In step 1, use 8 lb 10 oz raw ground beef. Continue with steps 2 and 3. In step 4, omit pinto or kidney beans. Continue with steps 5 - 7.

100 servings: In step 1, use 17 lb 4 oz raw ground beef. Continue with steps 2 and 3. In step 4, omit pinto or kidney beans. Continue with steps 5 -7.

Chili con Carne with Beans

Meat/Meat Alternate-Vegetable | Main Dishes | D-20

1 lb dry kidney beans = about 2 ½ cups dry or 6 ¼ cups cooked beans.

Nutrients Per Serving					
Calories	180	**Saturated Fat**	3.57 g	**Iron**	2.71 mg
Protein	15.44 g	**Cholesterol**	42 mg	**Calcium**	46 mg
Carbohydrate	10.68 g	**Vitamin A**	813 IU	**Sodium**	204 mg
Total Fat	8.58 g	**Vitamin C**	14.5 mg	**Dietary Fiber**	2.5 g

What are the vulnerabilities?

What restrictions can be applied?

Now, let's take the knowledge from this exercise and relate it to a functional food defense plan. Your food defense plan addresses all the vulnerabilities in your operation and then applies restrictions in the form of mitigation strategies and traceability to save lives and protect your brand.

A FUNCTIONAL FOOD DEFENSE PLAN

What Should Be in a Food Defense SHARE Plan?

In the development of your food defense plan, you will need to use the SHARE principles as you evaluate your overall facility. As a reminder, the SHARE principles are strategy, hazards, accessibility, restriction, and enforcement. The key is to compare and evaluate your written food defense plan against the actual systems in place and review your processes, behaviors, and human resources.

The key is to compare and evaluate your written food defense plan against the actual systems in place and review your processes, behaviors, and human resources.

The food defense plan is implemented when effective protective measures are identified and reinforced as intended. The operation needs to test or conduct a mock drill of your written food defense plan. This is essential. A written Food Defense SHARE plan should include the following five elements:

1. Vulnerability assessment
2. Mitigation strategies
3. Traceability
4. Verification
5. Recordkeeping and documentation

Let's explore each element in greater detail.

1. Vulnerability Assessment. It is critical to conduct a vulnerability assessment as your first step. This is the process used to identify weaknesses and top physical security priorities in your operation. The completed vulnerability assessments and self-inspection checklists from Chapters 1, 2, 3, and 4 are needed for your Food Defense SHARE plan. Complete column one of your food defense plan with the items marked as out of compliance according to your analysis.

2. Mitigation Strategies. The second step in a Food Defense SHARE plan is a description of corrective actions, or **countermeasures**, which are also known as mitigation strategies. Mitigation strategies are solutions for each area of weakness that was identified. Here is a list of potential mitigation strategies from the *Guidance for Industry: Retail Food Stores and Food Service Establishments: Food Security Preventive Measures Guidance* (*Contains Nonbinding Recommendations*—December 2003; Revised October 2007):

Mitigation strategies are solutions for each area of weakness that was identified.

- Watch for unusual or suspicious behavior by staff (e.g., staff who, without an identifiable purpose, stay unusually late after the end of their shift, arrive unusually early, access files/information/areas of the facility outside of the areas of their responsibility, remove documents from the facility, ask questions on sensitive subjects, bring cameras to work).
- Prevent access to food preparation, storage, and dishwashing areas in the nonpublic areas of the establishment, including loading docks.
- Monitor public areas, including entrances to public restrooms (e.g., using security guards, monitored video cameras, one-way and two-way windows, placement of employee workstations for optimum visibility) for unusual or suspicious activity (e.g., a customer returning a product to the shelf that he/she brought into the store, spending an unusual amount of time in one area of the store).

- Monitor the serving or display of foods in self-service areas (e.g., salad bars, condiments, open bulk containers, produce display areas, doughnut/bagel cases).
- Ensure that there is a valid reason for all visits to the nonpublic areas of the establishment before providing access to the facility. Beware of unsolicited visitors.
- Secure doors (including freight loading doors, when not in use and not being monitored, and emergency exits), windows, roof openings/hatches, vent openings, ventilation systems, utility rooms, ice manufacturing and storage rooms, loft areas, trailer bodies, and bulk storage tanks for liquids, solids, and compressed gases, to the extent possible (e.g., using locks, "jimmy plates," seals, alarms, intrusion detection sensors, guards, and monitored video surveillance [remember to consult any relevant federal, state, or local fire or occupational safety codes before making any changes]).
- Inspect incoming products and product returns for signs of tampering, contamination, or damage (e.g., abnormal powders, liquids, stains, or odors, evidence of resealing, compromised tamper-evident packaging) or "counterfeiting" (e.g., inappropriate or mismatched product identity, labeling, and product lot coding or specifications, absence of tamper-evident packaging when the label contains a tamper-evident notice), when appropriate.
- Reject suspect food.
- Alert appropriate law enforcement and public health authorities about evidence of tampering, "counterfeiting," or other malicious, criminal, or terrorist action.
- Monitor self-service areas (e.g., salad bars, condiments, open bulk containers, produce display areas, doughnut/bagel cases) for evidence of tampering or other malicious, criminal, or terrorist action.
- Establish a system of traceability of computer transactions.
- Know who is and who should be on premises, and where they should be located, for each shift.
- Keep information updated.

3. Traceability. The third step in a Food Defense SHARE plan is a description of a facility's procedures for tracing distribution. Distribution is traceability of your food products and/or ingredients from where you received them from and where they are going. In other words, this is one distribution level forward and one distribution level backward as required by the *Food and Drug Administration's (FDA) Reportable Food Registry (RFR) for Industry.* For more information, refer http://www.fda.gov/food/foodsafety/foodsafetyprograms/rfr/default.htm.

This information will be used in the event that your operation has come in contact with intentionally adulterated food. As part of having a functional Food Defense SHARE plan, use what you already have in place:

- List of approved suppliers and letters of guarantee from your HACCP food safety purchasing standard operating procedure
- Food recall standard operating procedure
- Employee contact list
- Disposal and decontamination procedure
- Crisis management team policies and procedures

The Food Defense SHARE Traceability Form is a sample that you can modify to achieve **supply chain communication** of suspect food products that may have entered the food supply. This form is located on page 66.

4. Verification. The fourth step in a Food Defense SHARE plan is a description of your verification procedures. Verification must be completed *annually* as part of your HACCP and Food Defense SHARE plans. Verification of your Food Defense SHARE plan is *required* when a food defense event or an alleged incident occurs. In addition, verification is necessary when there are

- key leadership changes, such as new personnel added to your operation;
- updates to regulations, like the FDA Food Code;
- changes in the flow of food, including significant modification at critical workstations; and
- new construction or renovations to the physical structure.

The same individual responsible for monitoring food defense self-inspections cannot complete the verification review. The Food Defense SHARE Verification Form is provided on page 68 for you to modify and customize to your operation.

5. Recordkeeping and Documentation. The last step in the Food Defense SHARE plan is a description of the facility's food defense recordkeeping procedures, which is similar to HACCP. Be sure to include a description of the facility's procedures for food recalls, whether those recalls are due to unintentional or intentional contamination. Employee health procedures are also essential for monitoring and following up with sick employees and absenteeism.

Following is a core list of food defense recordkeeping documents that should be used:

- Food Defense SHARE Plan
- Food defense SHARE Verification Form
- Food Defense SHARE Traceability Form
- Purchasing standard operating procedure
- Food recall standard operating procedure
- Employee contact list
- Disposal and decontamination procedure
- Crisis management team procedure
- Needs assessment
- Vulnerability assessment
- Self-inspection checklist

Now, It Is Time to Create a Food Defense SHARE Plan

Begin by completing the Food Defense SHARE Plan cover page. This cover page serves to provide general information as well as protect sensitive information regarding your operation.

Food Defense SHARE Plan

Cover Page

Establishment Name: ______________________

Establishment Location (city, state): ______________________

Preparer's Name: ______________________

Title: ______________________

Signature: ______________________

Date: ______________________

Product Description: ______________________

Method of Distribution: ______________________

Intended Use: ______________________

Consumer: ______________________

Strategy: This is setting the right tone to achieve food defense awareness throughout your operation. Refer back to page 3 in Chapter 1 to identify the vulnerabilities of your operation. If you answered No, then this is a vulnerability that should be listed in column one of your Food Defense SHARE Plan. Complete the form and write your food defense **S**HARE plan for Strategy.

	(1) Vulnerability Assessment	(2) Mitigation Strategies
Example	Food defense is not discussed or trained during the orientation process.	Use the food defense standard operating procedure for new hire orientation on pages 8–9.
1.		
2.		
3.		

Hazards: Hazards must be appropriately identified in order to prevent, eliminate, and reduce the risks associated with an intentional contamination. Refer back to page 15 in Chapter 2 to identify the vulnerabilities of your operation. If you answered No, then this is a vulnerability that should be listed in this section of your Food Defense SHARE Plan. Complete the form and write your food defense S**H**ARE plan for Hazard.

	(1) Vulnerability Assessment	(2) Mitigation Strategies
Example	Food defense procedures are not effectively communicated to all employees.	Have quarterly or monthly meetings and discuss food defense awareness principles, practices, and protocols for your operation.
4.		
5.		
6.		

Accessibility: Accessibility specifies access rights and control to specific areas of your operation. This is determined by whether you have direct control or influence. Control is your ability to directly change the behaviors of others within your operation. Influence is the power to leverage or sway the behaviors of others. Your control and influence were analyzed on pages 37–38 in Chapter 3, on Accessibility. If you answered "Control," without a specific name, or "Influence," then this is a vulnerability that should be addressed in your Food Defense SHARE Plan. Complete the form and write your food defense SH**A**RE plan for Authorization.

	(1) Vulnerability Assessment	(2) Mitigation Strategies
Example	All vendors do not have a food defense plan.	Require all vendors to have a food defense plan or replace the vendor.
7.		
8.		
9.		

Restriction: Restriction is based on imposing limitations and constraints throughout your operation following the flow of food. Refer to the self-inspection checklist located on pages 53–54. If you answered No, then this is a vulnerability that should be listed in this section of your Food Defense SHARE Plan. Complete the form and write your food defense SHARE plan for Restriction.

	(1) Vulnerability Assessment	(2) Mitigation Strategies
Example	Appropriate management personnel have not received training in food defense.	Train appropriate management personnel on food defense.
10.		
11.		
12.		

Restriction in Your Operation: Restriction imposes limitations and constraints throughout the flow of food. Complete this section of the Food Defense SHARE Plan that applies to your operation.

	(1) Vulnerability Assessment	(2) Mitigation Strategies
Purchase		
Receive		
Store		
Prepare		
Cook		
Hold		
Cool		
Reheat		
Serve		

Enforcement: Enforcement is the last step in Food Defense SHARE and will be addressed in greater detail in our next chapter. For the purpose of completing your Food Defense SHARE Plan, this activity has been deliberately placed in this section to emphasize the requirement for this section. This concludes the proactive and preventive aspects of the Food Defense SHARE Plan. Enforcement is important for your operation to respond appropriately during a potential contami-nation event. This is the step in your food defense plan when preventive measures were breached. Understand that enforcement requires you to activate your legal system due to a breach of protocol or a potential deliberate attack has been detected. Enforcement entails the process for investigating and detaining suspected food and/or a perpetrator. Complete the form and write your food defense SHARE plan for Enforcement.

	(1) Vulnerability Assessment	(2) Mitigation Strategies
Example	Employees are not trained to cooperate during an investigation.	Train employees to cooperate according to the policies and procedures.
13.		
14.		
15.		

Food Defense SHARE Traceability Form

Date: ______________________ Time: ______________

Individual Completing the Traceability Form: ______________

Phone: ______________

Cell phone: ______________

Text Acceptable: ______________

Yes or No ______________

E-mail: ______________

Fax: ______________

Address: ______________

Suspect Food Product ________	**Traceability One-Back Contact Information**	**Traceability One-Forward Contact Information**
Product Codes: Identifying marks on the label or container: Name of Company on the Product Label: Phone: E-mail: Address: Web site: ************************ ***************	Product: Co. Name: Phone: Date and time completed ________ Emergency Contact: Cell phone: Date and time completed ________ Text Acceptable: Yes or No If yes, date and time completed ________	Product: Co. Name: Phone: Date and time completed ________ Emergency Contact: Cell phone: Date and time completed ________ Text Acceptable: Yes or No If yes, date and time completed ________ E-mail: Date and time completed ________

Date of Purchase:

Name of Store where the Product Was Bought:

Phone:

E-mail:

Address:

Web site:

E-mail:

Date and time completed __________

Fax:

Date and time completed __________

Address:

Web site:

Fax:

Date and time completed __________

Address:

Web site:

Did you contact law enforcement?

Yes or No

If yes, date and time completed __________

Law Enforcement Contact:

Phone:

Report Number:

****

Was emergency treatment provided?

Yes or No - If yes, date and time of first treatment __________

Name of Hospital:

Name of Doctor:

Address:

Phone:

****

Attach a separate document with the names, addresses, and phone numbers of people affected.

Back-up Emergency Contact:

Cell phone:

Date and time completed __________

Text Acceptable:

Yes or No

If yes, date and time completed __________

E-mail:

Date and time completed __________

Back-up Emergency Contact:

Cell phone:

Date and time completed __________

Text Acceptable:

Yes or No

If yes, date and time completed __________

E-mail:

Date and time completed __________

Food Defense SHARE Verification Form

Did your operation experience any of the following? If yes, you need to update your Food Defense Plan

- Key leadership changes
- New personnel
- Updates to regulations
- New FDA Food Code
- Changes in the flow of food
- Modification at critical workstations
- New construction
- Renovations to the physical structure

Date of Annual Review	Person In Charge for Conducting Annual Review	Date of Traceability Exercise or Mock Food Recall Conducted
__/__/__		
__/__/__		
__/__/__		
__/__/__		
__/__/__		

Restriction Study Questions

1. What are the four sections of the self-inspection checklist?

   a. ______________________________

   ____

   ____

   ____

   b. ______________________________

   ____

   ____

   ____

   c. ______________________________

   ____

   ____

   ____

   d. ______________________________

   ____

   ____

   ____

2. What is the purpose of a food defense self-inspection?

3. What is the definition of restriction?

4. What are characteristics of vulnerability that decreases an attack?

5. What are the components in a Food Defense SHARE plan?

a.

b.

c.

d.

e.

APPLY YOUR KNOWLEDGE
(Close Hold Assignment)

1. What steps would you take to decrease potential vulnerabilities in your operation by providing restriction to employees, vendors, and customers?

2. How will you apply food defense restriction?

Congratulations! You have successfully completed *Restriction*, the fourth step in SHARE! Let's begin with Chapter 5, Enforcement. . . .

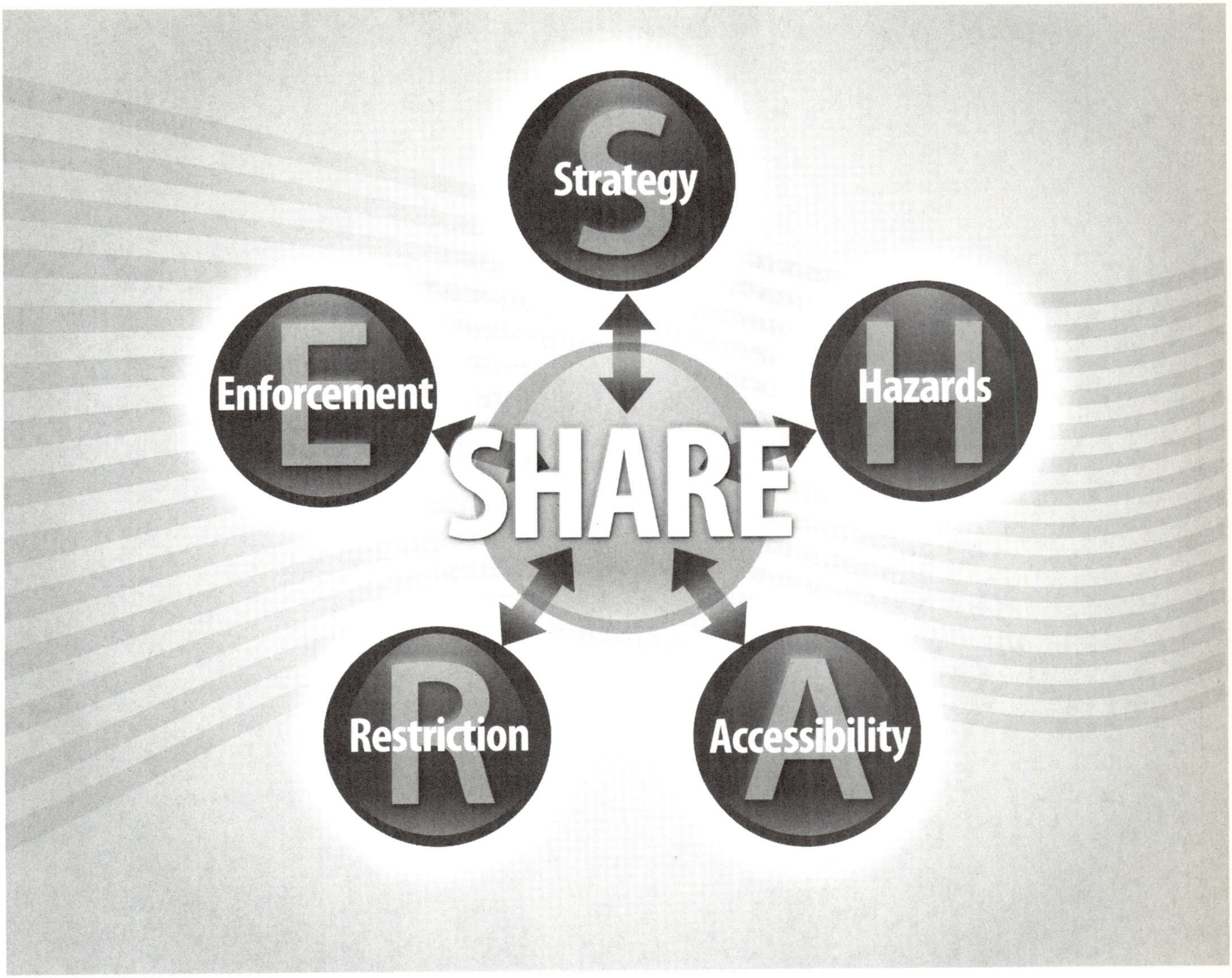

Chapter 5

Enforcement

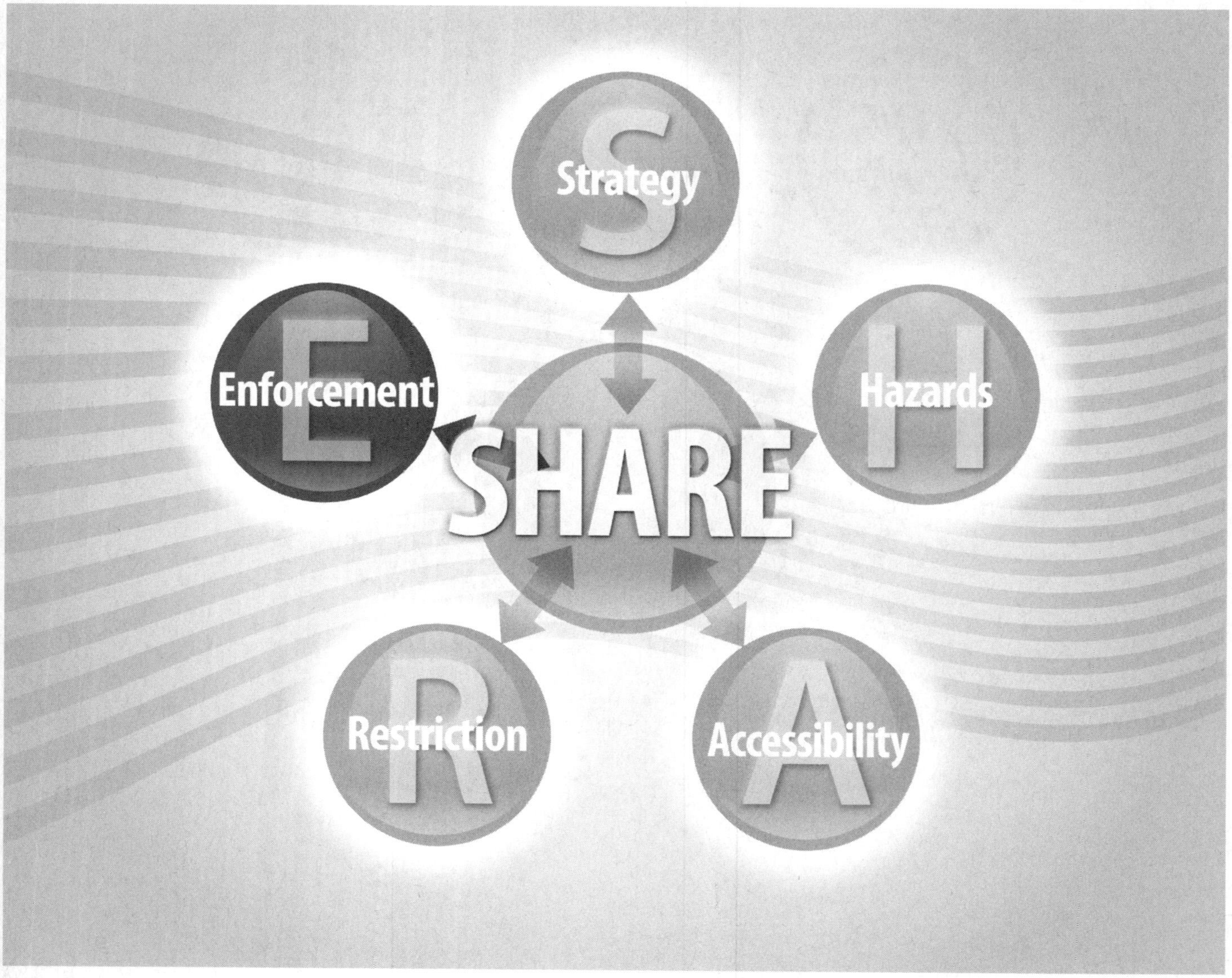

OVERVIEW

The goal of a Food Defense SHARE plan is to protect the food supply and your brand, which is not an option; it's a nonnegotiable commitment. In this chapter, we will no longer focus on prevention but will orient food defense thinking in terms of response. In Food Defense SHARE, *S*trategies, *H*azards, *A*ccessibility, and *R*estriction are the proactive preventive measures taken *prior* to an alleged incident. This chapter, *E*nforcement, is about how you can appropriately *respond to an alleged attack* on your facility.

Enforcement is traditionally defined as legal actions taken to gain compliance with laws, rules, regulations, and agreements resulting in penalties or criminal sanctions for violations. Enforcement is defined in Food Defense SHARE as actions taken to gain compliance by detection, monitoring, and reporting any breach in strategies, hazards, accessibility, and restriction.

Enforcement is traditionally defined as legal actions taken to gain compliance with laws, rules, regulations, and agreements resulting in penalties or criminal sanctions for violations.

In this final chapter, we will look at the many ways to assist law enforcement by knowing what is needed for an investigation and who is responsible for what role. Management needs to know their defined roles and responsibilities, which is to first take care of customers and employees and restore your operation from brand shock value in the aftermath of an attack. It is critical for you to be comfortable with all aspects of your food defense response: investigation, disposal, decontamination, prosecution, and recovery, collectively termed the "continuity of your business." A **Business Continuity and Disaster Preparedness Plan** means you have carefully considered your needs and understand what you need to know on how to recover and return all aspects of your business to the level of operation prior to the alleged incident. In Appendix 2, we have supplied the ready.gov Business Continuity and Disaster Preparedness Plan for you to complete and attach to your Food Defense SHARE plan—as the final component. Let's review the Food Defense SHARE goals for enforcement.

A **Business Continuity and Disaster Preparedness Plan** means you have carefully considered your needs and understand what you need to know on how to recover and return all aspects of your business to the level of operation prior to the alleged incident.

LEARNING OBJECTIVES

In this chapter, you will learn to:

- Understand enforcement as it relates to SHARE.
- Create written goals for preparedness, detection, emergency response, and recovery to be applied in your operation.
- Implement "If You See Something, Say Something" into your culture.
- Complete the Business Continuity and Disaster Preparedness Plan.

EXERCISE: WHAT IS ENFORCEMENT?

Let's analyze your operation as it relates to Food Defense SHARE and how you use enforcement. Enforcement is traditionally defined as legal actions taken to gain compliance with laws, rules, regulations, and agreements resulting in penalties or criminal sanctions for violations. Enforcement is defined in Food Defense SHARE as actions taken to compel compliance by detection, monitoring, and reporting any breach in strategies, hazards, accessibility, and restriction. In this exercise, we will focus on four operational goals for business continuity plan. These goals are preparedness, detection, emergency response, and recovery.

Enforcement is defined in Food Defense SHARE as actions taken to compel compliance by detection, monitoring, and reporting any breach in strategies, hazards, accessibility, and restriction.

What Are Your Goals?

Let's complete your enforcement goals independently; this is a close hold assignment and is better known as "top secret." Complete the following activity using the definitions provided below for preparedness, detection, emergency response, and recovery:

- **Preparedness** is defined as the state of being prepared, organized, and equipped for a specific purpose or event before hand.

FIGURE 5.1 A Food defense violation is a criminal act.

- **Detection** is to discover or ascertain the presence of a chemical or biological hazard using an ongoing systematic collection and analysis of data or close observation of a person or group.
- **Emergency response** is defined as addressing different emergencies by mitigating the destruction of human life or property, and ultimately ensuring public safety.
- **Recovery** is defined as restoring your operation from brand shock value in the aftermath of an attack, such as regaining what was lost or taken away, and returning back to normal, assuring consumers the products will be safe and secure following an event.

(A) Preparedness Goal:

____

____

____

(B) Detection Goal:

____

____

____

(C) Emergency Response Goal:

____

____

____

(D) Recovery Goal:

____

____

____

Exercise Summary

Enforcement helps you to think about the four goals necessary to gain compliance with laws, rules, regulations, and agreements resulting in penalties or criminal sanctions for violations. Enforcement is defined in Food Defense SHARE as actions taken to compel compliance by detection, monitoring, and reporting any breach in protocol. We hope that you take these goals back to your operation and SHARE them with your team as you implement food defense principles on a daily basis.

Reality Check

Secretary Napolitano and Administrator Johnson Announce Expansion of "If You See Something, Say Something" Campaign to Federal Buildings Throughout the Country

Release Date: December 8, 2010
For Immediate Release
Office of the Press Secretary
Contact: 202-282-8010

WASHINGTON—Department of Homeland Security Secretary Janet Napolitano today joined U.S. General Services Administration (GSA) Administrator Martha Johnson, Federal Protective Service (FPS) Director Eric Patterson, and GSA Commissioner of Public Buildings Bob Peck to announce the expansion of DHS' national "If You See Something, Say Something" public awareness campaign to approximately 9,000 federal buildings throughout the United States.

FIGURE 5.2 Food Defense involves everyone in the food ecosystem.

"Homeland security begins with hometown security and every citizen—including government employees—plays a critical role in ensuring America's safety and security," said Secretary Napolitano. "Our partnership with FPS and GSA to expand the 'If You See Something, Say Something' campaign to our nation's federal buildings is a crucial step in helping the millions of people who work in or visit our federal buildings every day identify and report suspicious activity indicators of terrorism, crime and other threats."

"GSA is pleased to partner with DHS to bring the 'If You See Something, Say Something' campaign to the over 9,000 properties we own or lease on behalf of the Federal Government," said GSA Administrator Martha Johnson. "The campaign is yet another important example of the work GSA and DHS do to ensure that our federal buildings are safe and welcoming to over one million federal employees and visitors alike."

The "If You See Something, Say Something" campaign—originally implemented by New York City's Metropolitan Transportation Authority and funded, in part, by $13 million from DHS' Transit Security Grant Program—is a simple and effective program to engage the public and key frontline employees to identify and report indicators of terrorism, crime and other threats to the proper transportation and law enforcement authorities.

(*continued*)

"The men and women of the Federal Protective Service work tirelessly every day to protect the safety of Americans who work in and visit federal facilities," said FPS Director Eric Patterson. "This collaborative security effort will enable our law enforcement and protective security officers to join forces with the public to secure federal facilities from threats."

In the coming weeks, "If You See Something, Say Something" public awareness materials designed to help America's businesses, communities and citizens remain vigilant and play an active role in keeping our homeland safe will be posted in approximately 9,000 federal buildings throughout the country. Signage will appear at FPS guard stations at each facility, and any calls reporting suspicious activity will be directed to the existing national network of FPS call centers, which operate 24 hours a day, seven days a week.

Over the past five months, DHS has worked with its federal, state, local and private sector partners, as well as the Department of Justice, to expand the "If You See Something, Say Something" campaign and the Nationwide Suspicious Activity Reporting (SAR) Initiative—an administration effort to train state and local law enforcement to recognize behaviors and indicators related to terrorism, crime and other threats; standardize how those observations are documented and analyzed; and expand and enhance the sharing of those reports with the Federal Bureau of Investigation and DHS—to communities throughout the country, including recent statewide expansions of the "If You See Something, Say Something" campaign across Minnesota and New Jersey. Other partners include Walmart, Mall of America, the American Hotel & Lodging Association, Amtrak, the Washington Metropolitan Area Transit Authority, the sports and general aviation industries, and state and local fusion centers across the country.

In the coming months, the Department will continue to expand the "If You See Something, Say Something" campaign nationally with public education materials and outreach tools designed to help America's businesses, communities and citizens remain vigilant and play an active role in keeping the country safe.

In the above article, the focus is on enforcement needed to decrease the impact of your brand shock value. Foodservice retail operators should use Food Defense SHARE in assisting law enforcement during an investigation. The concept of "If You See Something, Say Something" should empower you to take action in order to protect the global food supply and your brand. From the above-mentioned article, the one question that you must answer as part of this chapter is, "How would you apply 'If You See Something, Say Something' as part of your culture?" Remember, "If You See Something, Say Something" is a preventive measure you can use to expose vulnerabilities in your operation. Let's look at some key terms to understand enforcement and how it applies to your operation.

Key Term Match Game

Answer the questions below by filling in the blank with the appropriate word that best fits the sentence as it relates to the restriction protocols of your organization.

Enforcement	**Crime**
Justice	**Preparedness**
Continuity	**Detection**
Compliance	**Emergency Response**
Conviction	**Recovery**

1. ______________________________ is defined in Food Defense SHARE as actions taken to compel compliance by detection, monitoring, and reporting any breach in strategies, hazards, accessibility, and restriction.

2. ______________________________ is the quality of being fair, righteous, and lawful.

3. ______________________________ is the act of complying with a demand or request in accordance with established legislation, or guidelines.

4. ______________________________ is uninterrupted flow, or being continuous.

5. ______________________________ is defined as restoring your operation from brand shock value in the aftermath of an attack, such as regaining what was lost or taken away, and returning back to normal, assuring consumers the products will be safe and secure following an event.

6. ______________________________ is an evil act punishable by law.

7. ______________________________ is a judgment by a judge or jury in a court of law.

8. ______________________________ is defined as addressing different emergencies by mitigating the destruction of human life or property and ultimately ensuring public safety.

9. ______________________________ is to discover or ascertain the presence of a chemical or biological hazard using ongoing systematic collection and analysis of data or close observation of a person or group.

10. ______________________________ is defined as the state of being prepared, organized, and equipped for a specific purpose or event beforehand.

LEADERSHIP, SAFETY, AND SECURITY PROTOCOLS

What's Next? Where to Get started?

An individual that commits a harmful act, such as food terrorism, anywhere around the world is also known as an adversary, criminal, sociopath, terrorist, lone wolf, or the village idiot.

This chapter will address food defense as a criminal act. An individual that commits a harmful act, such as food terrorism, anywhere around the world is also known as an adversary, criminal, sociopath, terrorist, lone wolf, or the village idiot. These individuals oppose or act against laws, rules, regulations, and agreements of a governed society. They are focused on threatening our global food supply and livelihood. This "insider attacker" has a need to claim responsibility for an attack, disrupt commerce, or cause panic within our public health system. This person or group has a clear ideology, vision, and mission of causing harm and is not concerned about loss of his or her own life, nor the lives of other people.

As a rule of thumb, during an intentional food-related incident, the most prepared organizations would respond the best.

Within the global food chain, also called the farm-to-fork continuum, an attack may be first detected by a show of illnesses in animals and/or humans. As a rule of thumb, during an intentional food-related incident, the most prepared organizations would respond the best. Examples of such organizations include federal, state, tribal, and local regulatory agencies, as well as organizations from food industry sectors that can include farmers, distributors, manufacturers, processors, importers, transportation companies, and retail/foodservice operators such as convenience stores, supermarkets, restaurants, casinos, military, schools, senior care facilities, airports, institutional foodservice, cruise ships, contract foodservice, hotels, commissaries, hospitals, distribution centers, colleges, and universities. As a manager of any one of these organizations, it is critical to identify all of your vulnerabilities and apply mitigation strategies for the vulnerabilities identified. The second phase of enforcement is completing a Continuity and Disaster Preparedness Plan for your organization's response to an emergency.

From farm to fork, law enforcement works with the animal, plant, and food protection communities throughout the global food supply chain. In the United States, the leading organization for more information is the National Center for Food Protection and Defense (NCFPD). NCFPD's Web site is a resource for you to use as you prepare your Food Defense SHARE plan. It will assist your operation in addressing vulnerabilities to prevent, reduce, or eliminate the impact of an intentional contamination attack with biological or chemical agents. For more information, go to http://www.ncfpd.umn.edu/

Following are two examples of how biological or chemical agents are often first detected through medical symptoms, such as public health outbreaks, in the form of deaths or illnesses:

An animal death could be used as a warning sign that a terrorist is attempting to cause emotional and economic harm.

(1) From the farm, an animal population may display symptoms of severe sickness, as one of the first signs of an unintentional or intentional event. Farmers, transporters, processors, and veterinarians use these signs as an indicator or detection method before slaughter that there is a potential issue. Whether it is a pet or animal used for harvesting purposes, any pattern of illness or suspected foul play should be reported to law enforcement. The death of animals could be used as a warning sign that a terrorist is attempting to cause emotional and economic harm. The likelihood of sick cows making it to slaughter with **bovine spongiform encephalopathy** (BSE), commonly known as **mad cow disease**, would be unlikely as such cows would be rejected, based on international and federal regulations, laws, and guidance documents for processing sick animals such as downer cows.

(2) In the retail setting, if an incident occurs, management must determine or decide initially whether the event is/was an intentional or unintentional act. Traditionally, a manager's first instinct is to discard the food. This is a typical food safety response to accidental contamination of food. But in terms of food defense and law enforcement, you are required to think differently. Similar to a food safety incident, the food must be isolated and labeled "do-not-use." In the likelihood of deliberate adulteration, the manager must anticipate that the segregated food shall be used as evidence in the investigation.

Credibility of Evidence

The food industry needs to be aware of how the credibility of evidence assists law enforcement in making an arrest. Following are some indicators, behavioral factors, and examples that would assist public health and law enforcement professionals in determining the credibility of evidence:

- Respond to customer complaints, and suspicious activity, by keeping records and following up with threats and/or hoaxes.
- Investigate unusual inquires or excessive curiosity about obtaining samples, pictures, or asking questions.
- Report unexplained increases in sick and dying humans or animals.

Now that you have learned about the early warning signs for food safety or food defense events, the next step is to completely understand the criminal investigation process, as well as capturing critical information and supplying information based on your threat analysis. Let's begin with your criminal investigation process.

Criminal Investigation Process

Your role as a manager is to build relationships prior to an event. This will enable you to work more closely with the law enforcement community, local, tribal, state, and federal regulatory agencies. Your local Departments of Public Health and Centers for Disease Control and Prevention would also assist if there were an incident impacting public health. The public health activities would include surveillance, identification, communication, response, and damage control.

In most cases, a company will discuss all aspects of the event and have a clear message when law enforcement is called. This may include designating a spokesperson, retaining a subject matter expert, activating legal representation, and assisting in the investigation with law enforcement. Expect that the first responders will be law enforcement and will operate within the applicable law. These investigators will oversee proper collection of evidence and subsequently pursue prosecution. You can expect law enforcement to collect the following types of information:

- circumstantial evidence (supports the conclusion of fact);
- direct evidence (eyewitness statements);
- trace evidence (material residue);
- hearsay evidence (observations based upon what someone else told him or her); and
- eyewitness testimony (witness reporting smelling, tasting, or seeing something, or someone).

The role of law enforcement is to establish protocols necessary to conduct a threat analysis and determine the severity of the threat. If the threat is credible, law enforcement must take action to identify and prosecute violators.

Critical Information

To protect your establishment, managers need to know what is an inside attacker. The insider refers to an individual who has authorized access and a specific job function in your operation. These individuals have a legitimate right of entry to your facility and food products. Here is the key question each manager should ask, "If I were a criminal, what would I need to know to avoid detection and arrest, in order to carry-out an attack on my operation?" The insider's objective is to avoid detection, arrest, and gain your critical information needed to carry out a planned attack.

The insider refers to an individual who has authorized access and a specific job function in your operation.

Threat Analysis

Who would want to cause harm to your operation, brand, customers, and/or employees? As a manager you have a decision-making tool to use for food defense, which is your Food Defense SHARE plan. Threat agents for food defense are different than for food safety. Managers have been trained that biological, chemical, and physical contaminants are food safety hazards. In food defense, a physical hazard is not a hazard of concern because a terrorist seeks to hide contaminants and avoid detection. The terrorist does not want managers or employees to see, taste, feel, or smell anything unusual that would alert the person in charge of a potential problem. A threat analysis helps you to determine what critical information needs to be protected, such as access codes essential to your operation. For example, we need you to look more closely at your physical security, alarm detection, and standard protocols in your operation to verify its effectiveness.

The terrorist does not want managers or employees to see, taste, feel, or smell anything unusual that would alert the person in charge of a potential problem.

If the event is determined to be an intentional act, then law enforcement is activated and led by the Federal Bureau of Investigation (FBI). The next step for law enforcement is to obtain sufficient evidence and information to identify and then apprehend the individual or individuals responsible for the attack.

Law Enforcement Investigation Goals

Source: *Criminal Investigation Handbook for Agroterrorism*, July 2008.

Following are the investigation goals for law enforcement:

(1) Protect the public safety.
(2) Prevent a criminal act.
(3) Identify, apprehend, and prosecute the perpetrators.
(4) Protect law enforcement personnel.

Protect the Public Safety

Law enforcement role is to prevent an initial attack, apprehend a terrorist after an attack, and assist in preventing subsequent events from occurring. As a manager, your role is to help prevent an intentional act within your operation, communicate, and work with law enforcement.

Prevent a Criminal Act

Law enforcement's role is to obtain information that identifies potential terrorists, their targets, and methods of attack before an incident can be executed. A manager must focus on the policies, procedures, and management systems to protect your customers, employees, and brand. This is why "If You See Something, Say Something" must be a part of your culture.

A manager must focus on the policies, procedures, and management systems to protect your customers, employees, and brand.

Law enforcement needs you to safeguard the source of the intelligence information and the means by which it was gathered. This procedure avoids the inadvertent disclosure of sources and collection techniques, especially during ongoing operations. A manager needs to ensure that employees cooperate during all investigations.

Identify, Apprehend, and Prosecute the Perpetrators

Food defense SHARE allows managers and employees along with Law Enforcement to proactively engage in the fight of protecting the food supply against acts of food terrorism. The message to all employees should be "If You See Something, Say Something".

Protect Law Enforcement Personnel

In terms of food defense, a manager needs to know what a terrorist might use to adulterate the food supply and the hazards in your facility. Some managers around the country are participating in **tabletop** activities, vulnerability assessments, and exercises related to food defense. These role-play simulations help industry, regulators, and law enforcement respond to mock scenarios or incidences they may encounter. These exercises are designed to analyze the risk of exposure to a hazardous material or the threat agent.

Data Collection and Law Enforcement

The **collection of evidence** is varied and complex, such as from electronic motion detection on **closed-caption television (CCTV)** to searching through the establishment's trashcans for credible evidence. Other methods include (1) communication using scanners to monitor police radio networks or cellular phone conversations; (2) human observation via monitoring, listening, and asking questions; (3) imagery via still or video surveillance of personnel or the establishment; (4) open source using Web-based collection methods, and (5) using published information to profile the target.

FDA/USDA DECONTAMINATION AND DISPOSAL GUIDANCE TO INDUSTRY: CHARACTERISTICS OF AN INTENTIONAL ADULTERATION INCIDENT

FSIS and FDA are likely to become aware of an intentional adulteration incident by (1) receipt of a tip by a government agency (e.g., local law enforcement, FBI, USDA/Office of Inspector General) or the owner of the establishment; or (2) surveillance or monitoring system results (e.g., consumer complaint monitoring system, laboratory analytical results) indicating unusual patterns of foodborne disease incidence or product contamination.

Incidents involving the intentional adulteration of food products may differ from unintentional contamination incidents in a number of ways, including the following:

1. *Type of Agent*—threat agents may include biological, chemical, or radiological materials not typically found in food. Disposition of these unusual agents may require specific actions that provide special protection for personnel and limiting contamination of other environmental media. Actual remediation decisions will be made at the state and local level on a case-by-case basis in consultation with the impacted company and may differ from these suggestions.

- A Food Defense SHARE plan must designate who will respond to outside agencies about the hazards.

2. *Scale of the Incident*—although an intentional adulteration incident may be limited to a single facility, it could also involve a coordinated attack on a number of large facilities within a region or throughout the country. Larger-scale incidents may require coordination among local, state, and federal government agencies within a region as well as across regions and quickly overwhelm available resources.

- A Food Defense SHARE plan must construct a focused response script for dealing with outside organizations and agencies.

3. *Coordination with Local Law Enforcement and the FBI*—the law enforcement community will become involved upon the suspicion or confirmation of intentional contamination. FSIS and FDA staff in the field will need to work with state and local law enforcement and the FBI to rapidly retain or detain suspect products and take samples for laboratory testing while preserving the integrity of the crime scene. The ability of both of these types of actions to proceed unimpeded requires advanced planning among the different agencies involved.

The law enforcement community will become involved upon the suspicion or confirmation of intentional contamination.

- A Food Defense SHARE plan must designate who will work with law enforcement and/or state agencies.

4. *Public Concern*—media and the public, including government leaders, will have great interest and concern over an intentional contamination of the food supply.

Incidents of panic and confusion regarding which food products have been contaminated may occur. Effective stakeholder risk communications will be critical under these circumstances.

- A Food Defense SHARE plan must designate who will respond to outside agencies.

Procedures for Incidents Involving Intentional Adulteration

An intentional adulteration incident can be divided into several phases:

1. Discovery and Response Phase—the event is identified and confirmed, a response team is created, and the team conducts an investigation and detains, retains, and recalls contaminated products. Once the intentional adulteration is detected or even suspected, the appropriate authorities (FSIS, FDA, FBI, local law enforcement, state health authorities, etc.) should be notified.

- The need for a Food Defense SHARE plan must include these factors.

2. Disposal Phase—a specific plan for the disposal of the waste is developed and implemented. Oversight should be handled by an environmental quality authority to ensure that the contaminated products are disposed of properly and that there will be no further contamination of other media (i.e., air, groundwater, or soil).

- A Food Defense SHARE plan must designate who will respond to outside agencies.

Decontamination of facilities or equipment will likely occur after the contaminated products have been removed or collected and disposal response actions have been completed.

3. Decontamination Phase—affected equipment and structures are decontaminated based on a specific decontamination plan. Decontamination of facilities or equipment will likely occur after the contaminated products have been removed or collected and disposal response actions have been completed.

- A Food Defense SHARE plan must outline the steps for the continuity of operations.

4. Event Clearance and Termination—once the incident leader, or command representative, after consultation with all relevant agencies, declares the disposal and decontamination adequately completed, the incident is considered closed and terminated.

- A Food Defense SHARE plan must detail the execution of the continuity of operation procedures.

VOLUNTARY NATIONAL RETAIL FOOD REGULATORY PROGRAM STANDARDS—APRIL 2009—STANDARD 5

April 2009

Standard 5: Foodborne Illness and Food Defense Preparedness and Response

This standard applies to the surveillance, investigation, response, and subsequent review of alleged food-related incidents and emergencies, either unintentional or deliberate, which results in foodborne illness, food-related injury*, and outbreaks.

Requirement Summary

The program has an established system to detect, collect, investigate and respond to complaints and emergencies that involve foodborne illness, food-related injury*, and intentional and unintentional food contamination.

Description of Requirement

1. **Investigative Procedures**
   a. The program has written operating procedures for responding to and/or conducting investigations of foodborne illness and food-related injury*. The procedures clearly identify the roles, duties and responsibilities of program staff and how the program interacts with other relevant departments and agencies. The procedures may be contained in a single source document or in multiple documents.
   b. The program maintains contact lists for individuals, departments, and agencies that may be involved in the investigation of foodborne illness, food-related injury* or contamination of food.
   c. The program maintains a written operating procedure or a Memorandum of Understanding (MOU) with the appropriate epidemiological investigation program/department to conduct foodborne illness investigations and to report findings. The operating procedure or MOU clearly identifies the roles, duties and responsibilities of each party.
   d. The program maintains logs or databases for all complaints or referral reports from other sources alleging food-related illness, food-related injury* or intentional food contamination. The final disposition for each complaint is recorded in the log or database and is filed in or linked to the establishment record for retrieval purposes.
   e. Program procedures describe the disposition, action or follow-up and reporting required for each type of complaint or referral report.
   f. Program procedures require disposition, action or follow-up on each complaint or referral report alleging food-related illness or food-related injury* within 24 hours.
   g. The program has established procedures and guidance for collecting information on the suspect food's preparation, storage or handling during on-site investigations of food-related illness, food-related injury*, or outbreak investigations.
   h. Program procedures provide guidance for immediate notification of appropriate law enforcement agencies if at any time intentional food contamination is suspected.
   i. Program procedures provide guidance for the notification of appropriate state and/or federal agencies when a complaint involves a product that originated outside the agency's jurisdiction or has been shipped interstate.

2. **Reporting Procedures**
   a. Possible contributing factors to the food-related illness, food-related injury* or intentional food contamination are identified in each on-site investigation report.
   b. The program shares final reports of investigations with the state epidemiologist and reports of confirmed foodborne disease outbreak*s with CDC.

3. **Laboratory Support Documentation**
   a. The program has a letter of understanding, written procedures, contract or MOU acknowledging, that a laboratory(s) is willing and able to provide analytical support to the jurisdiction's food program. The documentation describes the type of biological, chemical, radiological contaminants or other food adulterants that can be identified by the laboratory. The laboratory support available includes the ability to conduct environmental sample analysis, food sample analysis and clinical sample analysis.

b. The program maintains a list of alternative laboratory contacts from which assistance could be sought in the event that a food-related emergency exceeds the capability of the primary support lab(s) listed in paragraph 3.a. This list should also identify potential sources of laboratory support such as FDA, USDA, CDC, or environmental laboratories for specific analysis that cannot be performed by the jurisdiction's primary laboratory(s).

4. Trace-back Procedures

a. Program management has an established procedure to address the trace-back of foods implicated in an illness, outbreak or intentional food contamination. The trace-back procedure provides for the coordinated involvement of all appropriate agencies and identifies a coordinator to guide the investigation. Trace-back reports are shared with all agencies involved and with CDC.

5. Recalls

a. Program management has an established procedure to address the recall of foods implicated in an illness, outbreak or intentional food contamination.

b. When the jurisdiction has the responsibility to request or monitor a product recall, written procedures equivalent to 21 CFR, Part 7 are followed.

c. Written policies and procedures exist for verifying the effectiveness of recall actions by firms (effectiveness checks) when requested by another agency.

6. Media Management

a. The program has a written policy or procedure that defines a protocol for providing information to the public regarding a foodborne illness outbreak or food safety emergency. The policy/procedure should address coordination and cooperation with other agencies involved in the investigation. A media person is designated in the protocol.

7. Data Review and Analysis

a. At least once per year, the program conducts a review of the data in the complaint log or database and the foodborne illness and food-related injury* investigations to identify trends and possible contributing factors that are most likely to cause foodborne illness or food-related injury*. These periodic reviews of foodborne illnesses may suggest a need for further investigations and may suggest steps for illness prevention.

b. The review is conducted with prevention in mind and focuses on, but is not limited to, the following:

1) Foodborne Disease Outbreaks*, Suspect Foodborne Outbreaks* and Confirmed Foodborne Disease Outbreaks* in a single establishment;
2) Foodborne Disease Outbreaks*, Suspect Foodborne Outbreaks* and Confirmed Disease Outbreaks* in the same establishment type;
3) Foodborne Disease Outbreaks*, Suspect Foodborne Outbreaks* and Confirmed Foodborne Disease Outbreaks* implicating the same food;
4) Foodborne Disease outbreaks*, Suspect Foodborne Outbreaks* and Confirmed Foodborne Disease Outbreaks* associated with similar food preparation processes;

5) Number of confirmed foodborne disease outbreaks*;
6) Number of foodborne disease outbreaks* and suspect foodborne disease outbreaks*;
7) Contributing factors most often identified;
8) Number of complaints involving real and alleged threats of intentional food contamination; and
9) Number of complaints involving the same agent and any complaints involving unusual agents when agents are identified.

c. In the event that there have been no food-related illness or food-related injury* outbreak investigations conducted during the twelve months prior to the data review and analysis, program management will plan and conduct a mock foodborne illness investigation to test program readiness. The mock investigation should simulate response to an actual confirmed foodborne disease outbreak* and include on-site inspection, sample collection, and analysis. A mock investigation must be completed at least once per year when no foodborne disease outbreak* investigations occur.

Outcome

A food regulatory program has a systematic approach for the detection, investigation, response, documentation, and analysis of alleged food-related incidents that involve illness, injury, unintentional, or deliberate food contamination.

Documentation

The quality records required to meet this standard include:

1. Logs or databases of alleged food-related illness and food-related injury* complaints maintained and current.
2. Collection forms specified in the operating procedures.
3. Investigation reports of alleged food-related illness, food-related injury*, or incidents. Reports are retrievable by implicated establishment name.
4. The written procedures, contracts or MOU's with the supporting laboratories.
5. The procedure addressing the trace-back of food products implicated in an illness, outbreak, or contamination event.
6. 21 CFR, Part 7, or written procedures equivalent to 21 CFR, Part 7 for recalls.
7. Completed copies of the annual review and analysis (after 12 months of data).
8. Current written media policy/procedure and contact person.
9. The contact list for communicating with all relevant agencies.
10. Portions of any emergency response relevant to food safety and food defense.

**Note*: See the Standards Definitions for the meaning of these defined terms. The time is now to complete your Food Defense SHARE Final Thoughts . . .

Final Thought Questions?

1. How will you restrict information as it relates to a need to know? Please explain.

2. What will you do if you suspect your food has been deliberately contaminate?

3. How will you get your employees to do the right thing and report what's wrong?

4. How can your organization apply "If you see something, say something."?

Enforcement Study Questions

1. What are the five types of evidence?

   a.

   b.

   c.

   d.

   e.

2. What are the four ways in which intentional and unintentional contamination differ?

   a.

   b.

   c.

   d.

3. What are the four phases in an intentional adulteration incident?

   a.

   b.

c. ______________________________

d. ______________________________

4. What are the four investigation goals for law enforcement?

a. ______________________________

b. ______________________________

c. ______________________________

d. ______________________________

5. Describe the five methods of data collection for law enforcement.

a. ______________________________

b. ______________________________

c. ______________________________

d. ______________________________

e. ______________________________

APPLY YOUR KNOWLEDGE
(Close Hold Assignment)

1. What concepts would you suggest to your organization to improve enforcement?

2. How will you apply food defense enforcement?

Congratulations! You have successfully completed the Food Defense SHARE Program; it is now time to take your Food Defense SHARE Certification Exam!

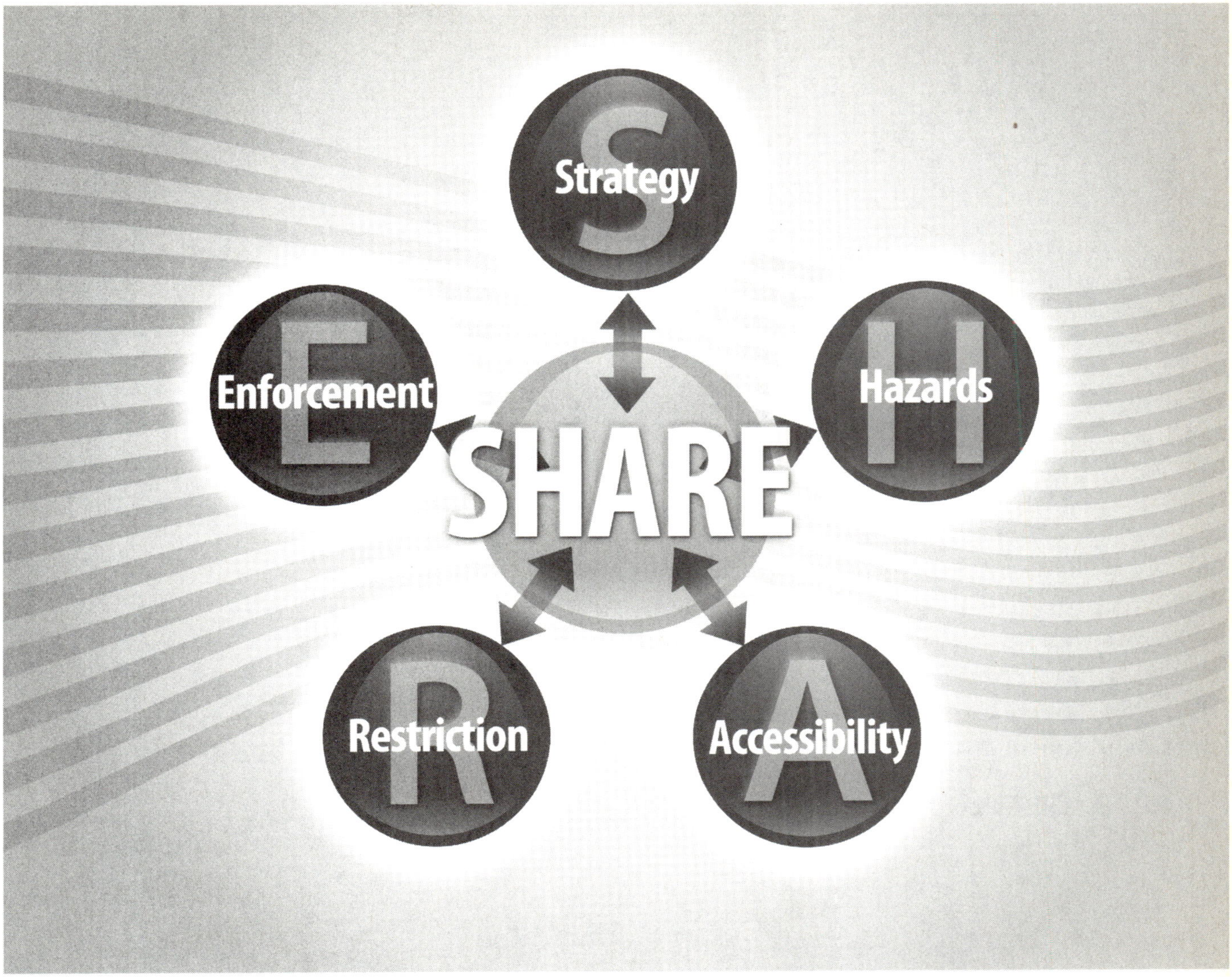

APPENDIX 1

Food Defense Plan
Security Measures for Food Defense

Establishment Name: ______________________________

Establishment Location (city, state): ______________________________

FSIS Establishment Number: ______________________________

By signing here, I acknowledge that this establishment has measures in place in accordance with this document.

Print Name: ______________________ Title: ______________________

Signature: ______________________ Date: ______________________

Food Defense Plan
Security Measures for Food Defense

__Food Defense__ is having measures in place to reduce the chances of someone intentionally contaminating the food supply in order to kill or hurt people, disrupt our economy, or ruin your business.

PURPOSE:

This voluntary plan documents your measures to protect food and food production processes from intentional harm. ***Review of this plan and signing the cover sheet will result in a Food Defense Plan for your FSIS-regulated establishment.***

BENEFITS: By having a Food Defense Plan, you will contribute to a safer and more secure food supply. You will also protect public health, your employees, and your livelihood. A functional* food defense plan may also:

- reduce the risk of unsafe product and economic loss,
- reduce theft,
- reduce the need for additional regulation on food defense, and
- reduce company liability.

INSTRUCTIONS:

1. Review the attached plan.
2. Sign the cover page.
3. On an annual basis, review this plan and document that you did so on the form in **Attachment B.**

This food defense plan is organized in four sections: (1) Outside Security Measures, (2) Inside Security Measures, (3) Personnel Security Measures, and (4) Incident Response Security Measures. **Attachment A** provides a list of tools or additional security measures that an establishment may consider or may already have in place. You may also have other plans that contribute to a food defense plan such as an emergency plan, a recall plan, a security plan, etc. **Attachment B** is a form that can be used to document your annual review of your food defense plan.

***The four elements that make up a *functional* food defense plan:**

1. **Develop:** Reviewing and signing this document fulfills this element.
2. **Implement:** Having measures described in this document fulfills this element.
3. **Test:** Periodic monitoring fulfills this element. This can be done using simple measures, such as checking locked doors or making unannounced perimeter checks. Monitoring can be documented using a form, such as **Attachment B.** Not all security measures need to be tested at the same frequency.
4. **Review and Maintain**: Reviewing the plan at least annually, revising the plan as needed, and taking appropriate actions fulfills this element.

This plan was developed through a joint effort between RTI International, the HACCP Consulting Group, a Focus Group of industry representatives, and the U.S. Department of Agriculture's Food Safety and Inspection Service.

Outside Security Measures

(Examples: door locks, lighting, monitoring loading/unloading)

GOAL: **To prevent unauthorized access by people, or entry of unapproved materials to the facility.**

This establishment has in place at least one of the following measures for outside security.

1.1 Physical Security

a. Plant boundaries are clear and secured to prevent unauthorized entry (for example, fences installed, no trespassing signs posted)
b. Entrances are secured (for example, locks and/or alarms installed and operating)
c. Plant perimeter is periodically monitored for suspicious activity
d. Outside lighting is present to deter unauthorized activities
e. Other access points such as windows and vents are secured
f. Outside storage on the premises is protected from unauthorized access
g. Other

1.2 Shipping/Receiving Security

a. Incoming shipments are examined for potential tampering
b. Incoming and outgoing vehicles are examined for suspicious activity
c. Loading and unloading activities are scheduled and/or monitored
d. Loading dock access is controlled (for example, monitored or locked)
e. Incoming shipments are secured with locks or seals
f. Outgoing shipments are locked or sealed
g. Other

1.3 Mail Handling Security

a. Mail is handled away from food including ingredients and packaged food product
b. Employees who handle mail are aware of proper handling of suspicious mail and U.S. Postal Service guidelines
c. Other

Inside Security Measures

(Examples: signs, observations, restricted access)

GOAL: To protect product from intentional contamination throughout the production process.

This establishment has in place at least one of the following measures for inside security.

2.1 General Inside Security

a. Suspicious packages are reported to appropriate personnel
b. Restricted areas of the establishment are clearly identified
c. Previously unattended materials are checked before use
d. Unexpected changes in inventory (product or equipment) are reported to appropriate personnel
e. Emergency lighting is in place
f. An emergency alert system is identifiable, tested, and reviewed with emergency contacts (for example, police or fire personnel)
g. Other

2.2 Slaughter/Processing Area Security

a. Access to live animals, ingredients, and packaged product is restricted
b. Access to animal handling areas and/or carcass coolers is controlled
c. Access to process control equipment such as ovens, mixers is restricted
d. Ingredients are examined for possible tampering
e. Records ensure traceability for one step backward, one step forward, or both
f. Other

2.3 Storage Security

a. Access to storage areas is restricted
b. Stock rotation (first in, first out) is practiced
c. Labels and packaging materials are controlled to prevent theft and misuse
d. Periodic examinations for tampering of materials in storage are performed
e. Other

Not all measures suggested are appropriate or necessary for every facility.

2.4 Ingredients/Water/Ice Security

a. Restrict access to storage tanks for potable water and to water reuse systems.
b. Access to lines that transfer water or ingredients are examined and restricted
c. Access to plant ice-making equipment is controlled
d. Restricted ingredients (for example, nitrites) are controlled
e. Supplier food safety/security information is requested
f. Other

2.5 Chemical/Hazardous Material Control Security

a. Chemicals/hazardous materials, including pesticides, cleaning or laboratory materials, and sanitizers, are in a restricted area or secured by a lock
b. Maintain an up-to-date inventory of hazardous materials and chemicals, and investigate discrepancies.
c. Potentially hazardous waste (biological or chemical) is controlled and disposed of properly
d. Other

2.6 Information Security

a. Access to sensitive information such as site plans and processing details is controlled
b. Access to computer systems is protected through firewalls and/or passwords
c. Other

Personnel Security Measures

(Examples: check references, use visitor log or sign-in, or check IDs)

GOAL: To ensure that only authorized personnel are in the facility at any time

This establishment has in place at least one of the following measures for personnel security.

3.1 Employee Security

a. A method to recognize or identify employees in the facility is in place

b. Background or reference checks are conducted for new hires[1]

c. Employees have restrictions on what they can bring in or take from the facility (for example, cameras)

d. Other

3.2 Non-employee Security (Example: visitors, contractors, guests, customers, truck drivers)

a. A log of non-employees entering the establishment is maintained

b. A method to recognize or identify non-employees in the establishment is in place

c. Non-employees are chaperoned on-site

d. Non-employees are restricted to appropriate areas

e. Non-employees have restrictions on what they can bring in or take from the facility

f. Other

3.3 Security Training

a. Awareness training on security measures is provided to new employees[2]

b. Refresher awareness training on security measures is offered to employees on a periodic basis

c. Employees are trained to report suspicious activities or unusual observations

d. Other

---

[1] You can electronically verify the employment eligibility of your new hires at http://www.dhs.gov/files/programs/gc_1185221678150.shtm. E-verify is an internet based system operated by the federal government that is available for employers to use at no charge

[2] You can access free food defense awareness training for your employees at http://www.fda.gov/Food/FoodDefense/Training/default.htm

Incident Response Security Measures

(Examples: reference your emergency plan, security plan or other)

GOAL: **To respond quickly to a product contamination threat or event using planned measures**

This establishment has in place at least one of the following measures for incident response security.

4.1 Investigating Security Concerns

a. Have procedures to ensure that adulterated or potentially harmful products are held.
b. Customer comments are investigated
c. Reporting unusual activities is encouraged
d. Information is available to employees on how to respond to phone or other threats
e. Employees have the ability to stop activities to minimize a potential food defense incident
f. Reported security breaches (for example, alarms, suspicion of tampering) are investigated
g. Other

4.2 Emergency Contact Security

a. Plant personnel contact information is kept up to date
b. Emergency contact lists are kept up to date
c. Other

4.3 Other Plan Security

a. A product recall plan is maintained and periodically reviewed
b. Key personnel are trained in product recall/withdraw procedures
c. Other

ATTACHMENT A

List of Tools or Possible Security Measures for Food Defense

This attachment provides a list of tools or additional security measures that an establishment may consider or may already have in place. These are provided to assist establishments in tailoring the plan to meet their specific needs.

1. Outside Security Tools

Physical Security

- Ensure proper lighting to monitor the establishment outdoors at night and early morning.
- Install self-locking doors and/or alarms on emergency exits.
- Ensure the following are secured with locks, seals, or sensors when unattended (after hours/weekends) to prevent unauthorized entry:
  - Outside doors and gates
  - Windows
  - Roof openings
  - Vent openings
  - Trailer (truck) bodies
  - Tanker truck hatches
  - Railcars
  - Bulk storage tanks/silos
  - Loading ports
  - Hose /Pump stations
- Regularly conduct and document security inspections of storage facilities, including temporary storage vehicles.
- Restrict outdoor access to water wells/sources.

Shipping / Receiving Security

- Closely monitor loading and unloading of vehicles transporting raw materials, finished products, or other materials used in food processing.
- Inspect tanker trucks and/or rail cars to detect the presence of any material, solid or liquid, in tanks prior to loading liquid products. Load only when appropriate. Report/record results.
- Control access to loading docks to avoid unverified or unauthorized deliveries.
- Require advance notification from suppliers for all deliveries.
- Immediately investigate suspicious changes in shipping documents.
- Check all deliveries against a roster of scheduled deliveries.
- Hold unscheduled deliveries outside establishment premises pending verification.
- If off-hour delivery is accepted, require prior notice of the delivery and an authorized person to be present to verify and receive the delivery.

Not all measures suggested are appropriate or necessary for all facilities.

- Check less-than-truckload (LTL) or partial load shipments for content and condition.
- Require incoming shipments of raw product, ingredients, and finished products to be sealed with tamper-evident or numbered, documented seals and verify the seals prior to entry. Reject if seals are broken or missing.
- Select transportation companies and suppliers with consideration of security measures that they use.
- Examine returned goods at a separate location for evidence of tampering before salvage or use in rework.
- Maintain records of disposition of returned goods.
- Require drivers or delivery personnel to provide identification, preferably with a photo ID. Record names.
- Minimize the time a truck is unlocked during loading or delivery.

2. Inside Security Tools

General Inside Security

- Install and monitor security cameras.
- Increase visibility within the establishment (for example, improve lighting, openness, increase supervision, add cameras).
- Regularly take inventory of keys to secured/sensitive areas of the establishment.
- Restrict access to controls (by locked door/gate or limiting access to designated employees) for the following systems:
    - Heating, ventilation, and air conditioning (HVAC)
    - Propane, natural gas, water, electricity
    - Disinfection systems
    - Clean-in place (CIP) systems or other centralized chemical systems

Slaughter / Processing Area Security

- Maintain records to allow efficient trace backward or forward of materials and finished product.
- Reduce the time an area is left unmonitored.
- Reduce access to product containers or processing equipment.
- Do not allow unnecessary personal items within the production area.

Storage Security

- Maintain an access log for product and ingredient storage areas.
- Regularly check the inventory of finished products for unexplained additions and withdrawals from existing stock.
- Restrict access to external storage facilities to designated employees only.

Ingredients / Water / Ice Security

- Examine packages of ingredients before use for evidence of tampering.
- Restrict access to product, ingredient, and packaging storage areas to designated employees only (for example, by locked door/gate).

Not all measures suggested are appropriate or necessary for all facilities.

- Water is from a municipally controlled source
- Inspect water lines for possible tampering (perform visual inspection for integrity of infrastructure, proper connections)
- Make arrangements with local health officials to ensure immediate notification to the establishment if the potability of the public water supply is compromised.

Chemical / Hazardous Material Control Security

- Restrict access to the in-plant laboratory.
- Have procedures in place to control receipt of samples.
- Have a procedure in place to receive, securely store, and dispose of reagents.

Information Security

- Track customer complaints/comments for trends.
- Keep details of food defense procedures confidential as necessary.
- Have up-to-date establishment layout/blueprints for local law enforcement, including the fire department if needed.

3. Personnel Security Tools

- Authorize appropriate employees to stop a process for significant concerns.
- Control access by employees and non-employees entering the establishment during working and non-working hours (use coded doors, receptionist on duty, swipe cards).
- Restrict temporary employees and non-employees to areas relevant to their work.
- Implement system to identify personnel with their specific functions, assignments or departments (for example, corresponding colored uniforms or hair covers).
- Prohibit employees from removing company-provided uniforms or protective gear from the premises.
- Maintain an updated shift roster for each shift.

4. Incident Response Tools

- Establish evacuation procedures and include in food defense plan.
- Establish procedures for responding to threats as well as actual product contamination events.
- Pre-establish communication with local, state, and federal incident response personnel for a more efficient response.

Not all measures suggested are appropriate or necessary for all facilities.

ATTACHMENT B

Food Defense Plan Review Form

Complete this form to document your annual review of this Food Defense Plan.

Not all measures are required or need to be reviewed each time this form is completed.

Date of Annual Review	Person Who Conducted Annual Review (Name and Title)	Was the Food Defense Plan tested?* (Yes / No)

*Testing can be done using simple measures, such as checking locked doors or making unannounced perimeter checks.

NOTE: Make as many copies of this attachment as necessary.

APPENDIX 2

Ready Business. **Business Emergency Plan**

Business Continuity and Disaster Preparedness Plan

☐ **PLAN TO STAY IN BUSINESS**

____
Business Name

____
Address

____
City, State, Zip Code

____
Telephone Number

The following person is our primary crisis manager and will serve as the company spokesperson in an emergency.

____
Primary Emergency Contact

____
Telephone Number

____
Alternative Number

____
E-mail

If this location is not accessible we will operate from location below:

____
Business Name

____
Address

____
City, State, Zip Code

____
Telephone Number

If the person is unable to manage the crisis, the person below will succeed in management:

____
Secondary Emergency Contact

____
Telephone Number

____
Alternative Number

____
E-mail

☐ **EMERGENCY CONTACT INFORMATION**

Dial 9-1-1 in an Emergency

____
Non-Emergency Police/Fire

____
Insurance Provider

For more information, visit www.ready.gov/business or call 1-800-BE-READY

Ready Business. **Business Emergency Plan**

Business Continuity and Disaster Preparedness Plan (cont'd)

☐ **PLAN TO STAY IN BUSINESS**

The following natural and man-made disasters could impact our business:

- ○ ______________________________
- ○ ______________________________
- ○ ______________________________
- ○ ______________________________

☐ **EMERGENCY PLANNING TEAM**

The following people will participate in emergency planning and crisis management.

- ○ ______________________________
- ○ ______________________________
- ○ ______________________________
- ○ ______________________________
- ○ ______________________________

☐ **WE PLAN TO COORDINATE WITH OTHERS**

The following people from neighboring businesses and our building management will participate on our emergency planning team.

- ○ ______________________________
- ○ ______________________________
- ○ ______________________________
- ○ ______________________________
- ○ ______________________________

☐ **OUR CRITICAL OPERATIONS**

The following is a prioritized list of our critical operations, staff and procedures we need to recover from a disaster.

Operation	Staff in Charge	Action Plan

Ready Business

Business Emergency Plan

Business Continuity and Disaster Preparedness Plan (cont'd)

☐ **SUPPLIERS AND CONTRACTORS**

Company Name: ______________________________
Street Address: ______________________________
City: __________ State: __________ Zip Code: __________
Phone: __________ Fax: __________ E-mail: __________
Contact Name: ____________________ Account Number: __________
Materials / Service Provided: ______________________________

If this company experiences a disaster, we will obtain supplies/materials from the following:

Company Name: ______________________________
Street Address: ______________________________
City: __________ State: __________ Zip Code: __________
Phone: __________ Fax: __________ E-mail: __________
Contact Name: ____________________ Account Number: __________
Materials / Service Provided: ______________________________

If this company experiences a disaster, we will obtain supplies/materials from the following:

Company Name: ______________________________
Street Address: ______________________________
City: __________ State: __________ Zip Code: __________
Phone: __________ Fax: __________ E-mail: __________
Contact Name: ____________________ Account Number: __________
Materials / Service Provided: ______________________________

Ready Business. **Business Emergency Plan**

Business Continuity and Disaster Preparedness Plan (cont'd)

☐ **EVACUATION PLAN FOR** ______________________________ **LOCATION**
(Insert Address)

The following natural and man-made disasters could impact our business:

- ○ We have developed these plans in collaboration with neighboring businesses and building owners to avoid confusion or gridlock
- ○ We have located, copied and posted building and site maps.
- ○ Exits are clearly marked.
- ○ We will practice evacuation procedures __________ times a year.

If we must leave the workplace quickly:

____

____

1. Warning System:______________________________

We will test the warning system and record results __________ times a year.

2. Assembly Site: ______________________________

3. Assembly Site Manager & Alternate:______________________________

a. Responsibilities Include:

____

____

____

4. Shut Down Manager & Alternate:______________________________

a. Responsibilities Include:

____

____

____

5. ____________________is responsible for issuing all clear.

Ready Business. **Business Emergency Plan**

Business Continuity and Disaster Preparedness Plan (cont'd)

☐ **SHELTER IN PLACE PLAN FOR** ______________________________ **LOCATION**
(Insert Address)

The following natural and man-made disasters could impact our business:

○ We have talked to co-workers about which emergency supplies, if any, the company will provide in the shelter location and which supplies individuals might consider keeping in a portable kit personalized for individual needs. ○ We have located, copied and posted building and site maps.

○ We will practice shelter procedures __________ times a year.

If we must take shelter quickly:

____

____

1. Warning System:__________

We will test the warning system and record results __________ times a year.

2. Storm Shelter Location:______________________________

3. "Seal the Room" Shelter Location: ______________________________

4. Shelter Location & Alternate :______________________________

a. Responsibilities Include:

____

____

____

5. Shut Down Manager & Alternate:______________________________

a. Responsibilities Include:

____

____

____

6. ____________________ is responsible for issuing all clear.

Ready Business. **Business Emergency Plan**

Business Continuity and Disaster Preparedness Plan (cont'd)

☐ COMMUNICATIONS

We will communicate our emergency plans with co-workers in the following way:

____

____

In the event of a disaster we will communicate with employees in the following way:

____

____

☐ CYBER SECURITY

To protect our computer hardware, we will:

____

To protect our computer software, we will:

____

If our computers are destroyed, we will use back-up computers at the following location:

____

☐ RECORDS BACK-UP

________________________ is responsible for backing up our critical records including payroll and accounting systems.

Back-up records including a copy of this plan, site maps, insurance policies, bank account records and computer back ups are stored onsite ____.

Another set of back-up records is stored at the following off-site location:

____

If our accounting and payroll records are destroyed, we will provide for continuity in the following ways:

____

Ready Business

Business Emergency Plan

Business Continuity and Disaster Preparedness Plan (cont'd)

☐ **EMPLOYEE EMERGENCY CONTACT INFORMATION**

The following is a list of our co-workers and their individual emergency contact information:

______________________ ______________________ ______________________

______________________ ______________________ ______________________

______________________ ______________________ ______________________

______________________ ______________________ ______________________

______________________ ______________________ ______________________

☐ **ANNUAL REVIEW**

We will review and update this business continuity and disaster plan in ______________________.

GLOSSARY

accessibility ability to physically access and egress (leave) from a target.

adulteration the adding or altering of a substance in a way that can cause harm to the person using or consuming it.

agent poisonous or toxic substances that could harm the environment, animals, and human beings. In human beings, these substances can cause death, illness, panic, and mistrust. Agents can increase the economic burden on industry and government.

background check a check done on a potential employee to determine issues in the individual's background that might impact his or her employment.

barriers obstacles that need to be managed.

benchmarking comparing your plan of action to the best practices in the industry on a continual basis.

biological hazards dangerous substances that are in the category of viruses, bacteria, parasite, fungi, and mold.

brand shock value (BSV) an expression used to measure the potential destruction of your brand or industry on a global scale.

Breach is a gap, infraction, or violation of standard operating procedures.

Business Continuity and Disaster Preparedness Plan plan that reflects careful consideration of needs and understanding to recover and return all aspects of your business to the level of operation prior to the alleged incident.

CARVER+Shock method a method that helps food processors protect their products from deliberate contamination.

chemical hazards dangerous substances in the category of cleaning agents, sanitizers, and pesticides.

close hold information not to be disclosed.

complex recipe a recipe that moves food through the temperature danger zone two times and is then prepared, cooled, stored, and reheated.

Compliance is the act of complying with a demand or request in accordance with established legislation, or guidelines.

concept a mental explanation or an abstract idea.

confidentiality agreement an agreement whereby a party agrees to keep certain information about a company or a business transaction confidential.

Constraints is the process for controlling and protecting critical information or areas.

Continuity is uninterrupted flow, or being continuous.

contractors an outside party that is contracted to provide services or products to a primary company.

control your ability to direct or change the behaviors of others in your operation.

Conviction is a judgement by a judge or jury in a court of law.

corrective actions predetermined steps that address weaknesses identified in a system.

countermeasures measures that are taken to prevent negative outcomes. In food defense, these may include enhancements to physical security, personnel security, and operational security to help minimize access to the product or process.

Crime is an evil act punishable by law.

criticality measurement of public health and economic impacts of an attack.

culture the particular characteristics of a group of individuals, an organization, or a civilization.

culture of commitment a culture in which every member of an organization is aligned in preventing a negative event, such as intentional contamination.

degree of openness the access to workstations and other systems.

dependency the ways that individuals depend on each other.

detection discovering or ascertaining the presence of a chemical or biological hazard using an ongoing systematic collection and analysis of data or close observation of a person or group.

documentation drafting, editing, formatting, submitting, reviewing, approving, distributing, and tracking.

Escherichia coli the abbreviation for *Escherichia coli,* a disease-causing bacteria found in the intestine of warm-blooded organisms. Certain strains can cause serious foodborne illness when contaminated food products are eaten.

effect amount of direct loss from an attack as measured by loss in production.

emergency response response that addresses different emergencies by mitigating the destruction of human life or property and ultimately ensuring public safety.

enforcement actions taken to compel compliance by detection, monitoring, and reporting any breach in strategies, hazards, accessibility, and restriction.

epidemiology the branch of medicine that studies the causes, distribution, and control of disease in specific populations.

flow of food the process steps in a foodservice operation that includes purchase, receive, store, preparation, cook, hold, cool, reheat, and serve.

Food and Drug Administration an agency of the U.S. Department of Health and Human Services responsible for protecting and promoting public health, including the supervision of health safety.

food defense the act of preventing an intentional contamination of food.

Food Defense SHARE plan A plan to protect food supply should include the five elements: vulnerability assessment, mitigation strategies, traceability, verification, and record keeping and documentation.

Guidelines is a formal principle or regulation.

hazard any biological, chemical, radiological, or physical agent that may cause a food to be unsafe for human consumption.

hazard analysis process used to assess risk of biological, chemical, and physical hazards throughout the flow of food.

Hazard Analysis and Critical Control Points (HACCP) HACCP is a proactive management system comprising seven principles in which food safety is addressed through the analysis and control of biological, chemical, and physical hazards from raw material production, procurement, and handling to manufacturing, distribution, and consumption of the finished product.

higher attractiveness higher vulnerability.

homeland security security efforts that protect the United States against terrorist activity.

human resources overall management of individuals, including hiring, training, disciplining, and terminating.

influence your ability to leverage or sway the behaviors of others within your operation.

information sharing the distribution of knowledge.

insider attacker a trusted individual who gains your approval and access with ease to cause physical or mental harm.

interdependency a system in which individuals depend on each other to complete various tasks, including those that have to do with food defense.

job description a document that describes the responsibilities and duties of a particular employee.

Justice is the quality of being fair, righteous, and lawful.

key workstations an assigned workspace or step within the flow of food.

leadership the ability to inspire others to achieve.

Limitations is keeping within certain limits.

management organizational activities that involve specific tasks such as planning, organization, and controlling to achieve goals.

Material Safety Data Sheets (MSDS) sheets that explain the details of the chemical and potential hazards.

medicine the study and treatment of health and body systems in human beings. Veterinary medicine is the study and treatment of health and body systems in animals.

metrics methods of measuring.

microbiology the study of *microorganisms*, which are unicellular or microscopic organisms.

mitigate to make less severe or painful.

mitigation strategies strategies that can help minimize potentially dangerous situations, including those that have to do with food contamination.

motive something that causes a person to act.

National Center for Food Protection and Defense Collaborators the organization charged with leading the food and agriculture sector in developing a multidisciplinary and action-oriented research consortium focused on identifying vulnerabilities in the nation's food system.

need to know selective sharing of information.

No-cook/simple recipe a recipe that does not move through the temperature danger zone. There is no cooking, or kill step, involved.

nodes the accessible parts of an operation.

nondisclosure agreement see confidentiality agreement.

nonnegotiable a situation that cannot be argued or negotiated.

Norovirus a virus that causes over 90 percent of viral outbreaks of gastroenteritis around the world. Norovirus is spread through food contaminated by feces, person-to-person contact, and contamination of surface.

observations conclusions that may involve a range of methods: informal interviews, direct observation, discussions, or analysis of documents.

open the most vulnerable and accessible area in your operation.

openness the degree to which a workstation or process is accessible to others.

parameter a boundary, limit, characteristic, or factor.

Physical Hazards foreign objects such as dirt, glass, fingernails, and metal fragments.

plan of action an ongoing performance analysis of your proce-dures, practices, and processes.

practices guidelines, customary or habitual behaviors, to perform in the most efficient manner.

preparedness the state of being prepared, organized, and equipped for a specific purpose or event beforehand.

prerequisite program standard operating procedure that addresses basic operational conditions in an establishment.

preventive controls control systems that prevent negative events, such as food contamination.

procedures a particular course of action to achieve a desired result.

processes systematic series of coordinated tasks and activities to achieve a desired outcome.

proprietary information information that is not public or general knowledge. The recipient of this information is required to keep the information private and to not use it in unauthorized ways.

protection guarding against or reducing risk.

protocol code of conduct, set of rules, and customs regarding behaviors.

psyche forces that influence thought and behavior.

Radiological Hazards are loose particles of radioactive material, such as plutonium and uranium.

recognizability in CARVER, the ease of identifying a target or a node.

records documents that demonstrate verification activities, such as vulnerability assessments, baseline physical security surveys, a written food defense plan, results of ongoing verification, or documentation from the reassessments.

recovery restoring an operation from brand shock value in the aftermath of an attack, such as regaining what was lost or taken away, and returning back to normal; assuring consumers the products will be safe and secure following an event.

recuperability in CARVER, the ability of a production system to recover.

resiliency the ability to prepare for, respond appropriately to, and recover from disasters.

Restrict is restricting access and actions.

restriction restriction focuses on having a well-defined physical security protocol and related activities that include job duties for oversight, monitoring, and surveillance by key personnel within the boundaries of your operation.

Right to Know a law that requires employers to tell employees the hazards one is exposed to in the workplace.

risk assessment the determination of risk related to a situation and a hazard.

Rules is a course of action.

Salmonella a disease-causing bacteria that can be transferred between humans and animals and can cause serious foodborne illness when contaminated food products are eaten.

same-day recipe a recipe that means the product is prepared for same-day service or has some same-day cooking involved, where the food will pass through the temperature danger zone one time before it is served or sold.

sector a distinct area.

self-inspection program a program that can help reduce the vulnerability of your food establishment to terrorist threats and criminal acts of product tampering.

share a food defense system of food protection that includes strategy, hazards, authorization, restrictions, and enforcement.

Shock a combination of health, economics, and psychological impacts of an attack.

standard operating procedure the appropriate chronological steps to follow within your operation.

stealth behavior that includes acting in a covert or sneaky manner.

strategy a plan of action.

subsystems in food defense, any added ingredients, water, premixes, and so on.

supportive culture a culture in which members are supported in their mission, including food defense.

susceptibility is a weakness or sensitivity allowing an action to be performed. This can be a characteristic that makes the global food supply vulnerable.

toxic agent any material or substance that is poisonous to human beings, animals, or the environment.

toxicology a branch of biology, chemistry, and medicine that studies the negative impact of chemicals on living organisms.

traceability the ability to track food from farm to fork.

training the teaching of skills and knowledge that develops knowledge, skills, and abilities.

United States Department of Agriculture (USDA) the federal department that administers programs that provide services to farmers (including research and soil conservation and efforts to stabilize the farming economy); created in 1862.

validation a preventive control that, when properly implemented, is capable of providing mitigation strategies geared to a specified outcome.

verification activities that ensure that the food defense plan is properly implemented.

vigilance the act of paying close attention to and watchfulness of your surroundings.

vulnerability in CARVER, the ease of accomplishing a desired outcome.

vulnerability assessment tool used in food defense to simply analyze the entire operation for weaknesses.

worst-case scenario a situation in which everything that could occur negatively has occurred.

zero tolerance automatic punishment for any violations of a rule, with the intention of immediately eliminating undesirable conduct.

INDEX

A

accessibility, 31
  control of, 51
  determining, factors for, 70
  food defense culture, requirements of, 50
  levels of, 50–54
  management of, 51
  reality check, application of, 54
  as shared responsibility, 56–58
  systematic approach to, 51
  types of, 51–54
accessibility rights, 46
acetone, 49
adulteration. *See also* contamination
  corrective actions for, 36
  economic, 50
  food defense culture for, 18
  intentional, 105–106
adversaries, 102
agents, 18, 102, 105
airports, 24–25
al Qaeda, 21
ammonia, 49
arsenic, 48
assumptions, 31

B

background check, 8
barriers of food defense/safety, 2
biological agents, 102
biological hazards, 18, 31, 35
Bioterrorism Act of 2002, 7
bleach, 49
bovine spongiform encephalopathy (BSE), 102
branding of food, 35
brand shock value (BSV), 64, 64–65
BSE. *See* bovine spongiform encephalopathy (BSE)
BSV. *See* brand shock value (BSV)
Business Continuity and Disaster Preparedness Plan, 97
business continuity plan, operational goals for, 97–98

C

CARVER (criticality, accessibility, recuprability, vulnerability, effect, and recognizability)+Shock methodology, 24–26. *See also* individual components of acronym
  attributes of, 31
  brand shock values in, 68
  conducting, steps for, 31–32
  function of, 30, 33–34
  vulnerability assessment of, 31
  weaknesses of, 32
casinos, 26
CCTV. *See* closed-caption television (CCTV)
CDC. *See* Centers for Disease Control and Prevention (CDC)
Centers for Disease Control and Prevention (CDC), 22, 103
channel count comparison, 29
chemical agents, 102
chemical awareness training, 9–10
chemical hazards, 18, 31
circumstantial evidence, 103
closed accessibility, 53
closed-caption television (CCTV), 105
close hold assignments (top-secret), 2, 64, 97
collection of evidence, 105
colleges, 26
commissaries, 29
commitment, culture of, 2
confidentiality agreements, 8
Consumer Self-Service Operations (C), 32
contamination. *See also* adulteration
  deliberate, 24, 45
  food defense culture for, 19
  of global food supply, 23
  of ground beef, 4–5
  industrialization as opportunity for increased, 24
  intentional, 2, 5, 102
  of spinach, 24
  unintentional, 2, 5, 24, 105
continual self-inspection, 6
Continuity and Disaster Preparedness Plan, 102
continuous monitoring, 35
contractors, 6
control, 37, 45, 50
convenience stores, 28–29
corrective actions, 36, 84
countermeasures for food safety, 6, 31, 84
Criminal Investigation Handbook for Agroterrorism, 104
criminal investigation process, 103
criminals, 102
critical information, 103
criticality, 31
critical limits in food defense culture, 34
cruise ships, 25
culture, 2, 9. *See also* food defense culture; food safety culture
curry, 48
customer awareness SOP, 9
cyanide, 21, 48

D

decontamination, 105–106
deliberate contamination, 24, 41
Department of Homeland Security (DHS), 9, 21, 99
detection, 97
DHHS. *See* New Hampshire Department of Health and Human Services (DHHS)
DHS. *See* Department of Homeland Security (DHS)
direct control, 45
direct evidence, 103
discovery and response team, 106
disposal, 105–106
Division of Public Health Services (DPHS), 66–67
documentation procedures, 6, 8, 86
  for Hazard Analysis and Critical Control Point, 38
domestic terrorists, 18
DPHS. *See* Division of Public Health Services (DPHS)

E

E. coli, 4
easy accessibility, 52
economic adulteration, 51
effects, 31
emergency response, 97
employee awareness SOP, 9
employee's food defense audit protocol, 11
enforcement
  defined, 97–98
  of food defense plan, 104

enforcement (*Continued*)
goals of, 97
law, 104–105
epidemiology, 31
evidence, 102–103
expert team, 31
eyewitness testimony, 103

F

facility awareness SOP, 9
farm-to-fork continuum, 102.
See also global food supply/chain
FDA. *See* Food and Drug Administration (FDA)
FDA *Food Code,* 32, 51–53, 86
Federal Bureau of Investigation (FBI), 100, 104
first in, first out (FIFO) food product rotation, 53
flow of food, 12, 20
critical issues within, 30
defined, 31
hazards in, 18
mitigation strategies for, application of, 64, 70
restriction to, 64, 94
vulnerability to, 64
food and agriculture (FA) sector, 30
Food and Drug Administration (FDA), 7, 33, 85, 105–106
foodborne illness, 23
in airports, 24–25
in casinos, 26
in colleges, 26
in commissaries, 29
in convenience stores, 28–29
on cruise ships, 25
within government, 29
in grocery stores, 29
in hospitals, 27–28
in hotels, 26
impact of, public health and economic, 24
in military restaurants, 29
in schools, 27
in senior care facilities, 28
in universities, 26
food defense audit protocols, 10–11
food defense awareness, 69
food defense critical control points, 34
food defense culture
accessibility, requirements of, 47
for adulteration, 18
agents, for biological and chemical, 18
barriers of, 2
chemical awareness training in, 10
for contamination, 18
critical limits in, 34
defined, 8
food safety culture and, goals linking, 33
in global food supply/chain, 23–24
ideal, 10
impact of, 24
importance of, 12
initiatives for, 6
for intentional contamination, 2, 5
leadership in, 5, 7
management's role in creating, 5
monitoring of, 10
precision machine in, incorporation of, 2
SOP for, 8
understanding, 23
vulnerability assessment factors in, 20
Food Defense Mitigation Strategies Database, 32
Food Defense SHARE (strategy, hazards, accessibility, restriction, and enforcement) plan, 2, 66. *See also* Hazard Analysis and Critical Control Point (HACCP); individual components of acronym
close hold assignments within, 2, 18
corrective actions of, 36
creating, 86–91
culture for, development of, 9–10
development of, 84
goal of, 56
implementation of, 84–85
importance of, 18
leadership in, responsibilities of, 5
management of, 4
mishandling, 2
as proprietary information, 2
self-inspection and, 6
strategies of, evaluating, 6–7
training of, 4
vulnerability assessments of, 2, 18, 33
written steps in, 7–8, 84–86
Food Defense SHARE Traceability Form, 86
Food & Drug Safety Program, 65
Food Protection Inspectors, 66
food safety culture
barriers of, 2
countermeasures for, 6
and food defense, goals linking, 33
for unintentional contamination, 2, 5
food terrorism, 4, 5, 102
foreign terrorists, 18

G

global food supply/chain, 23–24, 102
globalization, 23
government, 29
grapes, 48
grocery stores, 29
ground beef, 4, 48
Guidance for Industry: Retail Food Stores and Food Service Establishments: Food Security Preventive Measures Guidance, 84

H

HACCP. *See* Hazard Analysis and Critical Control Point (HACCP)
HACCP Food Safety Training Manual, 8
HACCP Star, 23, 32–33
Hartsfield-Jackson International Airport (Atlanta), 25
Hazard Analysis and Critical Control Point (HACCP), 5.
See also Food Defense SHARE (strategy, hazards, accessibility, restriction, and enforcement) plan
conducting, 33–34
corrective actions of, 36–37
critical limits of, 34
defined, 23
documentation of, 38
of food defense critical control points, 34
importance of, 18
monitoring, 34–36
philosophy of, 23
physical hazards, addressed by, 18
prerequisite program for, 23
as proactive management system, 5
record keeping of, 38
self-inspection and, 6
understanding, 23
verification for food defense and, 37
hazards, 18, 31, 33, 35
hearsay evidence, 103
high accessibility, 52

hospitals, 27
hotels, 26
human resource food defense audit protocol, 10
human resource systems, 8, 10–11
human tragedy, 51

I

"If You See Something, Say Something" public awareness campaign, 9, 99–100, 104
incentives, 35
interdependency, culture of, 2
industrialization, 24
influence, direct, 45
information sharing, 5
infrastructure, vulnerabilities within, 20
insiders, 103
intentional adulteration, 102, 105–106
intentional contamination, 2, 5, 102
intermittent monitoring, 35

J

Japan tsunami, 18
Johnson, Martha, 99

L

law enforcement, 104–105
leadership, 5, 8, 64
lone wolf, 102
low accessibility, 53

M

management
 of accessibility, 50
 criminal acts, role in preventing, 104
 food defense culture, role in creating, 5
 of Food Defense SHARE plan, 84
medium accessibility, 52–53
Metropolitan Transportation Authority, New York City, 9, 99
microbiology, 31
military restaurants, 29
mitigation strategies, 7, 84
 for flow of food, 64, 70
 restrictions, for applying, 70, 84
 for vulnerabilities, 32
"mom-and-pop" stores, 28
monitoring food defense culture, 10, 34–36
Montero, José, 66
motives, 50, 54

N

NAKAM, 48
Napolitano, Janet, 99
National Center for Food Protection and Defense (NCFPD), 38, 102
National School Lunch Program (NSLP), 4, 27–28
Nationwide Suspicious Activity Reporting (SAR) Initiative, 100
NCFPD. *See* National Center for Food Protection and Defense (NCFPD)
New Hampshire Department of Health and Human Services (DHHS), 66–67
New Hampshire Department of Justice Consumer Protection, 67
nicotine-based insecticide, 48
nodes, 30–31
non-accessibility, 53–54
noncontinuous monitoring, 35
nondisclosure agreements, 8
Norovirus outbreak, 25
NSLP. *See* National School Lunch Program (NSLP)

O

observations within food defense culture, 10
one-store operations, 29
on-site self-inspection, 6
open accessibility, 51–52

P

parameters, 31
Patterson, Eric, 99–100
Peck, Bob, 99
physical hazards, 18
physical security food defense audit protocol, 11
practices, defined, 51
precision machine, 2
preparedness, 97
prerequisite program, 23
proactive management system of HACCP, 5
procedures, 35, 51
processes, 35, 51
product tampering, 66
proprietary information, 2
psyche, 4
public concern, 105
public safety protection, 104

Q

qualified leader responsibilities, 8

R

radiological contamination, 18
radiological hazards, 18
Rajneeshee cult, 48
rat poison, 48
reality check, 54
recognizability, 31
recordkeeping procedures, 8, 86
 for Hazard Analysis and Critical Control Point, 38
 for monitoring food defense culture, 10
recovery, 98
recuperability, 31
Reportable Food Registry (RFR) for Industry (FDA), 7, 85
resiliency, 5, 64, 70
restrictions
 defined, 64–65
 to flow of food, 64, 68
 mitigation strategies for applying, 70, 84
 vulnerabilities, for defending, 64
rewards, 57
ricin, 18
"Right to Know" law, 10
risk assessment, 31
rodenticide, 49

S

salads, 48
Salmonella, 4, 35, 48
sanitation controls, 37
SAR. *See* Nationwide Suspicious Activity Reporting (SAR) Initiative
schools, 27
self-inspection, 64
 checklist for, 64
 continual, 6
 in Food Defense SHARE plan, 6
 in HACCP, 6
 importance of, 66
 on-site, 6
 vulnerability, for reducing, 66
senior care facilities, 28

SHARE. *See* Food Defense SHARE (strategy, hazards, accessibility, restriction, and enforcement) plan
shared responsibility, 56–58
sociopaths, 102
SOP. *See* standard operating procedure (SOP)
spinach, 24
standard operating procedure (SOP), 8–9
stealth, 52
strategies
  culture of commitment, requirement of, 2
  of Food Defense SHARE plan, evaluating, 7–8
  importance of, 12
  plan of action, as nonnegotiable component of, 2, 5
  purpose of, 2, 5
  requirement of, 2, 5
  role of, 2
  strength of, 2
Strategy Vulnerability Assessment, 3
strict liability, 24
supply chain communication, 85
supportive culture for removing barriers to improve food defense, 2
susceptibility, 30

T

tabletop activities, 104
temperature danger zone, 69–70
terrorists, 18, 102, 104
threat analysis, 103
time limitations, 52–53
toxicology, 31
traceability, 7, 33, 84–86
trace evidence, 103

U

unintentional contamination, 2, 5, 24, 105
U.S. Census Bureau, 28
U.S. Department of Agriculture (USDA), 4, 105–106
universities, 21
USDA. *See* U.S. Department of Agriculture (USDA)

V

VA. *See* vulnerability assessment (VA)
validation, 37
vendor awareness SOP, 9
verification procedures, 7, 37, 85
village idiot, 102
voluntary national retail food regulatory program, 106–109
vulnerabilities, 31
  characteristics of, 69
  countermeasures for, 31
  defined, 18, 69
  determining, factors for, 69
  to flow of food, 64
  food defense, 20
  within global food supply/chain, 24
  within infrastructure, 21
  nodes of, 31
  restrictions, for defending, 64
  self-inspection, for reducing, 66
  understanding, 20–21
vulnerability assessment (VA), 2, 7, 84
  of CARVER+Shock methodology, 32
  defined, 20–21
  food defense culture, factors in, 20, 64
  of Food Defense SHARE plan, 2, 18, 33
  hazard, 20

W

wine, 48
World Health Organization, 50
worst-case scenarios, 30–31